Tim,

We need more .\
Tully's se?\
All the b\
of success.

MW01484115

HAUNTING
THE CEO

A tale of
true leadership
in an era of
IT failure

John D. Hughes

Published by Spotless Books
1700 Seventh Ave., Suite 2100
Seattle, WA 98101

The advice and strategies within this publication may not be suitable in your
situation. Consult with an appropriate professional for your specific situation.

Hughes, John D., 1962–

Haunting the CEO : a tale of true leadership in an era of IT failure / John D. Hughes.

ISBN: 978-0-615-35600-6

Library of Congress Control Number: 2010906243

Printed in the United States of America

First Edition

10 9 8 7 6 5 4 3 2 1

For my parents, Jack Hughes and Vickie Peterson Hughes.
I miss you both more than you could ever know.
You awakened the words that follow.

Acknowledgements

The list is endless.

Thank you to those who took the time to read the manuscript when it was simply a Word document; to my many friends who only got to read the first three chapters and were then left hanging for several months (sorry!); to the baristas who brewed my tea (hot and iced) just right; and to each of you who inspired me along the way with your encouraging and positive words. Your efforts and gestures are deeply appreciated.

Lynda McDaniel—you are the best friend a writer ever had, thank you for your skilled guidance. Barb Rowan—thank you for your brilliant design abilities. And Larry Murray, your proofreading expertise is gratefully appreciated. (If any typos remain, keep in mind that a Red Sox fan proofread the book!) And Susan Snyder, thank you for the encouragement only one writer could give another.

Finally, to Aaron, Jessica and Laura, thank you for your help with the book, for being the best kids (and now adults) God ever created, and for your patience and understanding with your dad. I love you. And speaking of God, thank you for your faithfulness and for the gift and joy of writing.

Contents

Introduction

This is a book about leadership.

Leadership isn't so much about action, but about people and heart and desire and growth and being pulled forward to something better and more right than today. It's also about invisibility. No, not invincibility—*invisibility*.

When the best leader's work is done the people say, "We did it ourselves." —Lao Tzu

I wrote *Haunting the CEO* with CEOs and CIOs in mind. It works well too if you desire to be one of these, or just want to become a better leader wherever you're situated in business or life. Jump in.

Information Technology (IT) organizations are simply too often broken. They not only don't meet the needs of the business, but they distract and disrupt the business, turning critical capital dollars into costs and lost opportunities intended for business advantage.

I am convinced that leadership alone is the answer to broken IT organizations.

True leadership can bridge the communication chasm that

has for too long existed between business and IT, and true leadership can also bring about greater business growth, profitability and innovation. This book points the way.

I grew up in IT, starting with my first job as a Computer Operator at the age of 19 just before I graduated with my two-year degree in computer programming. I soon became a high-tech COBOL programmer. (Similar to Etch a Sketch© being a high-tech drawing device.)

But instead of coding, I kept sticking my head up above the cubicle walls to see what else was going on around me. I couldn't stay in the detail of code. I wasn't so much bored as I was curious.

For whatever reason, I paid attention. Mostly to leaders, or at least those in leadership positions. I watched, listened and learned from them, whether I reported to them or not. I remember the good ones, the bad ones, the bad ones who occasionally did good, and the good ones who occasionally did bad. I paid attention and learned from every one of them.

What you'll read in this book comprises nearly 30 years of paying attention. I stored many of those paying-attention moments—the events, feelings, expressions, epiphanies, successes, failures and characters—in my mental leadership database. A few of these moments appear in this story. Enjoy them, get mad at them, love them, hate them, laugh at them, disagree with them, question them, use them, forget them, throw the book at them if you must… but enjoy them nonetheless.

The three core leadership traits you'll read about came out of these handful of years. I developed them through observation, but also through my own trials and errors. They are road tested. Develop them within yourself and you'll understand true leadership.

Most of the stories in this book have a tinge of truth to them. And the characters are a mish-mash of real-life characters I've known. Names are made up to protect the innocent, and the not-so-innocent. But know that I'm the biggest not-so-innocent in the book. My mistakes are woven throughout the stories, characters and events you're about to immerse yourself in. Feel free to learn from my failures and missteps. I don't mind.

What's the point? IT has always frustrated me. When people ask me why I started my own CIO consulting business I always tell them that I want to fix every broken IT organization in the country. Most are still laughing. But that vision pulls me forward anyway. I soon learned that I can't accomplish this by myself. So I'm asking for your help. Whether you're a CEO, a CIO or reside somewhere in between, this book is my way of enlisting your help. Thank you for doing your part.

John D. Hughes
August 12, 2010
Woodinville, WA

Only three things happen naturally in organizations: friction, confusion, and underperformance. Everything else requires leadership.

—Peter Drucker

Prologue

Jim Cavo sat behind his new desk at Cantril Distribution. It was Friday evening, late, and everyone had gone home. He wasn't scheduled to officially begin until Monday, but he wanted to get a better feel for what he was getting himself into. He opened his desk drawer and pulled out last year's financial results. He rubbed the back of his neck as he scanned the figures for the fifth time. He didn't like it, but he knew he was going to have to make some changes.

Across town, Brian Kagey was helping his wife, Jill, make dinner. Helping wasn't quite the right word. He'd burned the garlic toast and over-cooked the spaghetti, lost in worry about the new CEO coming on board that week. He'd been through this before. A new leader usually meant a new team. He was a damn good technologist, and he could play most any role in IT. It just wouldn't be fair if he was let go.

Carol Lee was already missing Jim at Keoslin Industries. He'd been a good CEO there, and she wished he hadn't moved on. But she understood his drive, his desire to always be making a bigger difference. Cantril would be a big challenge for him, and his competitive side just couldn't pass up the opportunity. Still, she was glad he'd called. The fact that he thought she could help his new company made her feel like she did something right as his CIO at Keoslin.

1

Fear Renewed

"You wanted to see me, Jim?"

I had politely rapped on the new CEO's door and attempted to utter those words confidently, but instead they crawled out of my mouth a crackling, nervous whisper. Not a good first impression.

"Yes I did, Brian. Close the door and have a seat."

The strength and confidence in his voice shook me. I walked into his office and sat in one of the two leather chairs in front of his large oak desk. Although the chair was comfortable, it didn't relieve my concern for what I was about to hear. I hadn't met with Jim since he joined us on Monday. I'd been in a couple meetings with him and saw him around the building, but this would be our first one-on-one discussion.

Jim didn't hesitate. "I've worked for several companies over the last 30 some years. Throughout that time, I have only experienced one IT organization that performed well, that met the real needs of the business. I've seen IT disrupt businesses, fail to deliver on critical projects and put technology ahead of customer and business needs. And..."

Jim's phone rang. He apologized and said he needed to take it and that he would only be a minute.

I nervously looked around, trying to take my mind off where this conversation was headed. Several moving boxes lined the wall to my right. From a quick glance I saw books, pictures and several autographed baseballs in clear plastic cases. I wondered if he'd played in college or the majors, or maybe he was just a collector.

He didn't look much like an athlete. He was older, thin and very distinguished looking, not the worn and rugged look you might expect from an ex-ballplayer. He spoke confidently, but with a reserved tone.

His desk was bare, with the exception of a small wood-framed picture, a laptop and a printed document with a silver pen resting on top of it. I assumed the picture frame contained a photo of his family, but it faced him, so I could only guess.

From my interactions with him, Jim wasn't the kind of leader who was charismatic or loud or demanded attention. He was actually very unassuming. This thought relaxed me a bit, until I noticed that his laptop was open. There wasn't a docking station or monitor on his desk, and I worried that maybe the request to order and install them fell through the cracks, or worse, that it wasn't submitted at all. "Damn," I said to myself, trying to imagine what Patrick's excuse might be this time for the help desk messing up again. As usual, I'd have to take care of it myself.

Still trying to keep my mind occupied, I peered again at the document on Jim's desk. It looked familiar. *It was my résumé!* My heart shot into my throat, my stomach twisted. My time as CIO at Cantril Distribution was over. The feeling inside me was all too familiar. I had been here before. This would end my third CIO role in barely five years. I had failed again. Damn. What was wrong with me?

As these debilitating thoughts banged inside my head, I glanced up to see that Jim had already ended his phone conversation. He was looking at me as if I'd just been talking out loud, to no one. He began speaking without acknowledging my obvious worry.

"Sorry about that interruption," Jim said. "As I was saying, most of the IT organizations I've known were able to meet only the basic technology needs of the business, and maybe one helped improve internal productivity. But none helped drive greater business growth and profitability... until my last company."

I decided to make a pre-emptive strike. I mustered as much confidence as I could, but my voice still quivered. "We have a pretty good IT team, Jim. I've been here about a year-and-a-half, and we've been able to stabilize some technology problems the company was experiencing." As soon as the words left my mouth, I felt like I'd spoken out of turn.

Jim didn't seem to care; he simply continued as if I hadn't said a word.

"But at Keoslin Industries, my previous firm, our IT organization actually helped drive business growth and profitability. In fact, I would go so far as to say they helped us innovate. I need that here, Brian."

I desperately wanted to say okay and confidently express that I knew how to make these things happen. I didn't. Instead, my words fell flat on their face. "Okay, Jim." My stomach knotted more tightly and my heart sank even deeper.

Still unaffected by my interruptions, Jim continued. "I must admit that until my last company, I didn't like IT. I came to view it as a necessary evil, something we had to throw millions of dollars into just to be in the game. It felt like I couldn't ever expect a return on those dollars... that didn't seem to be part of the deal.

"The IT organizations I'd been exposed to previously weren't able to deliver business value, no matter how much technical talent they possessed. They were like black holes sucking in capital and cash we desperately could have used elsewhere in the organization. I learned to not trust IT. But Keoslin changed that. I should say that my CIO there changed my perception of IT. I've seen it happen. I know that IT *can* help drive greater business growth, profitability and even innovation."

Crap. I didn't know how to respond. I'm a technology leader, not a business leader. Sensing where this discussion was headed, I slumped lower in my chair and prepared myself for the worst.

And Jim confirmed it. He picked up my résumé, glanced at it, looked me in the eyes and said "Brian, we need to make some changes."

2

Haunting the CEO

Jim's words sucker-punched me; my wife's face flashed before me. I felt inadequate, a failure. How was I going to explain to her that this was happening again? If I can't keep a job as a CIO, what will I do? How would I support my family?

Then I remembered my comment to Jim about my IT organization being "pretty good." Based on that comment, I'm sure he saw through me, saw that I really didn't know how to help the business grow and innovate. To make matters worse, even I didn't think my IT team was that good. "We can't even get a docking station and monitor on the CEO's desk!" I said to myself. Damn it. Just damn it.

Jim clearly saw the fear and hurt in my eyes. But this time he acknowledged it with a slight, almost wry, smile.

"Brian, I was SVP of Marketing at a large hospitality firm about ten years ago. It was September and the executive management team was offsite at our annual strategic planning session. During one particular meeting, we were taking turns presenting drafts of our individual plans and answering the CEO's questions. We also took the opportunity to ask him questions.

"When it was our CIO's turn, he looked the CEO dead in the eyes and asked him in all seriousness, 'What keeps you awake at night?' The CEO's face hardened and without hesitation, he responded *'You do!'*"

"The CIO was taken aback by that response. Truthfully, we all were. He looked as if he'd just been betrayed by his best friend. He had no clue that he wasn't meeting the CEO's expectations. Brian, he managed IT, but he didn't lead IT. He operated as someone who managed a technology group, not as someone who was trying to grow a business. The CEO saw the problem, but the CIO didn't, and that was his downfall."

I wanted to interrupt, but it was clear I would only hurt my case at this point. I decided to stop interrupting and just listen.

"Brian, managers move a company from day to day, but they don't move the needle in strategic, profitable ways. Don't get me wrong, we need managers, good ones. But to move the needle in big ways we need leaders."

I continued to keep my mouth shut. This was an interesting story, but I still didn't see where he was headed or how this affected me.

Jim continued. "Our CEO went on to explain to us that he felt like IT haunted him in his dreams. It haunted him in his office. It haunted him in his car. He didn't have the transparency he needed from IT. He didn't trust IT. He told us that at times he did lay awake wondering if tomorrow would be the day he saw his company's name in the Wall Street Journal detailing a major information security breach. Or he worried about the latest multi-million-dollar IT project that was three months late, 50 percent over budget and no end in sight. He wasn't getting value out of IT. He had learned not to trust IT, and ultimately that meant he couldn't trust the CIO."

The knot in my gut tightened as Jim finished his story. I didn't know what was about to happen; I just knew I couldn't take this gut-wrenching feeling any longer. It felt like my career was in ICU, and I was in the waiting room wondering if it would survive.

"Brian, I'm going to be replacing some of my direct reports and other senior executives. I've learned to quickly assess leadership strength and potential, and the bottom line is Cantril has a tremendous opportunity to strengthen its leadership talent. These are strong managers, able to deliver operationally, but I need top executives who can grow a company, not just operate one. It will be a painful transition for everyone involved, but a necessary one. I've already spoken with the individuals affected by my decisions.

"And that brings me to you, Brian. You were a bit tough for me to assess. Your IT organization is nowhere near what I need, so my decision about your future with Cantril was based on potential. We haven't interacted with each other much, but I did get plenty of feedback about you from your peers, your team and others throughout the company. Based on this feedback and the traits I desire in leaders, I've made my decision.

"Brian, I'm going to stick with you, for now. You still have to deliver, but you'll get the chance to learn how."

I didn't know whether to jump for joy because I still had a job, or cry tears of sorrow because I had no clue how to meet Jim's high expectations for IT. Either way, the anguish would continue. The taste of bittersweet victory roiled within me.

"Thank you, sir," was all I could muster. Pathetic. Like Oliver Twist lifting his empty bowl and begging for more porridge.

"Please sir, can I have some more?"

I looked down and noticed that I had been tightly clenching my chair's seat with both hands. For a second I didn't think I'd be able to let go, I was so tense. But I wanted to hear all that Jim had to say. I wanted to learn why he was keeping me and what he had in store for me. So, I calmed myself down and slowly released the death grip on the chair.

Even Jim appeared more relaxed now. With a bit more up-beat tone in his voice, he continued. "I mentioned earlier that we needed to make some changes. Actually, there's only one change you need to make, and two changes I need to make.

"Brian, I need you to meet with Carol Lee. Carol was my CIO at Keoslin Industries. She will be able to provide more detail about how she operates as a CIO, and what she does to produce business results through IT. Don't be afraid to ask questions or challenge her. She's always open to learning and improving herself and her IT organization. I'll have her contact you."

"Okay," I agreed, though with some reluctance, unsure if I would be meeting with Carol to be helped or assessed further.

"As for me," Jim continued, as if on a roll now with his thoughts, "I'm going to make two changes. First, I'm moving IT out from under the CFO. You now report directly to me, Brian."

"Okay," I said, attempting to hide my nervousness about re-porting directly to Jim. After all, I liked Tom. As CFOs go, he was easy to work with, although he was always pressuring me to cut costs to the bone and keep headcount lower than I wanted.

Jim continued. "HR will also report directly to me. IT and HR are too critical to our success to push down into the organization.

"My second change is easier said than done. I now expect more out of IT. Significantly more. I'll be holding you to a higher degree of accountability than you've experienced here. And because of our situation, I need to begin seeing improvements in IT almost immediately. You get a chance, Brian, but the runway won't be long."

"It sounds like I'll have to rebuild the airplane during take-off!"

Thankfully, Jim laughed. "That's what I need from you, Brian. Cantril's revenues have flattened over the last three years, profits are down 30 percent and we've lost market share. To get these numbers moving in the right direction, I need greater results from every area of the company, especially IT.

"I need a return on the dollars invested in systems and technology. I expect your organization to help produce greater business growth, profitability and eventually innovation. We'll see what you can learn from Carol to quickly move us in this direction."

Jim and I finished our conversation. As I left his office, my emotions were churning inside me. I still had my job, but no idea how to accomplish what Jim wanted. And Carol would certainly recognize this.

It felt like I was teetering on the edge of a cliff. I was certain that I would fall. If not now, eventually. The good news was that Jim's request for me to meet with Carol would buy me time to look for another job. I wasn't going to wait around to fail... or be fired.

3

Too Fast

"Hi Brian. My name is Carol Lee. Jim Cavo wanted me to call and schedule a time for us to meet."

That was fast. 8:15 a.m. the next morning. Jim must think I need a lot of help. I tried to sound nonchalant. "Hi Carol. Yes, Jim mentioned something about wanting me to meet with you."

"Is there a time tomorrow that you could get away from the office for a couple hours for a coffee?"

I really wasn't ready for this discussion. I was hoping for more time to kick off my job search. I decided to stall. "Um, no, not really. I'm sure you're as busy as I am Carol, so could we schedule an hour next Thursday or Friday afternoon?"

"I have plenty of time to meet tomorrow, Brian," Carol said, sounding relaxed. "I think it's important for us to get together this week. I can drive to the Eastside to save you some time. We could meet at the Starbucks down the street from your office building."

"Could we make it next Wednesday?" I countered. I had a full day of meetings the next two days that I needed to sit in on. We had a warehouse application project kicking off, I needed to attend my data center operations staff meeting, and there was no way I could let Kris, one of my business analysts, meet with the SVP of Sales by herself; she's just not ready.

"I think it's important that we meet this week, Brian. Are you sure you don't have a couple hours available?"

I flashed back to my meeting with Jim. It felt like he was hovering over my shoulder, observing my interaction with Carol. I didn't want to risk pushing back again. "Okay. I can meet at 8:30 Thursday morning."

"And we'll meet until 10:30, okay?" Carol added.

I didn't like giving in, but there it was. "Sure," I said. Now I'd have to cancel the project kick-off meeting and reschedule it for next week, which would put the project a week behind before it even starts. I would also have to trust that the infrastructure team could stay focused without me, which was iffy, and hope that my business analysts wouldn't make promises to the sales team that we couldn't keep.

Carol described herself so that I could recognize her. The word "plain" would have sufficed, I thought. She also shared that she would be wearing a pink blouse and black slacks. I said that I would likely wear a blue shirt and blue suit pants. I usually did. We exchanged cell numbers and said goodbye. I slammed the phone down.

This was moving too fast. I shoved my hands into my hair, curled my fingers and pulled until it hurt. I didn't know how to give Jim what he wanted out of IT, and I had little time to figure it out. Meeting with Carol wasn't going to help. What could she possibly know about being a CIO that would save my job, my career and help Jim grow this company? I had failed again. I pulled my hair harder… and prayed for a miracle.

4

CIO Seizure

The warm aroma of coffee brewed from freshly ground beans welcomed me into the café. I took in a deep breath and held it for a second. Heaven should smell this good. I scanned the waiting line and seating area but didn't see anyone that fit Carol's description of herself. I had arrived a few minutes early anyway. I just couldn't afford to make a bad first impression. Carol might be the only person standing between me and a résumé upgrade.

I pulled out my phone and started reading through the dozen emails I'd received since leaving my office 10 minutes ago. I started to respond to the first one when I heard my name.

"Brian?"

I spun around to find an attractive woman in her mid-to-late forties standing in front of me. Not the Plain Jane she had described. Her outfit stylish but professional, her straight dark hair pulled behind her ears, her dark-rimmed glasses perfectly framed her face.

I nodded. "Yes. Are you Carol?"

She reached out her hand. "I am. It's nice to meet you, Brian."

I reach for her hand. "It's nice to meet you too, Carol." My voice cracked as I said her name.

She leaned her head toward the counter. "Would you like some caffeine?"

I smiled nervously. "Sure, you can never have enough caffeine in your veins." Carol managed a tiny laugh. I appreciated the effort.

She had a warm smile and a kind voice. I was expecting a tougher, more assertive personality given that she was the CIO of a billion-dollar company. I was pleasantly surprised and a little disarmed. I exhaled, realizing I'd been holding my breath.

"I'll have a double tall mocha." The words rolled off my tongue like I'd ordered that same drink every day for the last seven years… which was about right.

"And I'll have a tall Earl Grey tea," Carol said to the barista.

Carol paid and we picked up our drinks. An older couple had just freed up a small round table in the corner of the café, so we grabbed it. "The corner will give us a little more privacy," Carol said.

We spent some time talking about our careers. I shared with her that I was in my third CIO role, and that I still enjoyed working in the details of networking and server administration. We took a couple tangents into the personal topics of family, college and our favorite places to visit in the Pacific Northwest. Carol eventually moved us to the reason for our meeting.

"Brian, before we start, I want you to know that you have total freedom to speak your mind. I won't share any of our conversations with Jim or anyone else, unless you give me permission. You have my complete confidence. Besides, Jim doesn't care about the details of our conversations. He's not laying in wait to judge you. He only wants results."

Those were odd initial comments. I wasn't quite sure why she said them, but I guess I didn't quite care. I was too stunned by her use of the plural "conversations."

"You mean we'll meet more than just this one time?"

Carol laughed politely. "Well, I suppose that's up to you, Brian. Let's see how our first discussion goes, and then you tell me if you want to continue or not."

I had that sinking feeling that for me to have a chance at keeping my job, Carol and I were going to become best buddies.

"Okay. That's fair," I said.

Carol smiled. "Good. I know Jim pretty well, and I doubt that he shared much with you about why we're meeting."

I stared Carol in the eyes and tapped the side of my nose with my right index finger, hoping Carol would get the reference to charades and the signal for "That's right, you got it!" Thankfully, she laughed and continued.

"Brian, my husband started his own accounting business several years ago. Luckily, a friend gave him a book called *The E-Myth*. Although I wasn't going to be involved in the business, I wanted to support him in any way I could, so I also read the book. And I'm glad I did."

"Does the "E" stand for electronic?' I asked.

"No. It stands for 'entrepreneur.' The book talks about the different hats a small business owner has to wear, and that trying to wear all of those hats will actually limit the growth and profitability of his business. They wear a technician's hat, a manager's hat, and an entrepreneur or leader's hat."

Carol continued. "Michael Gerber, the author of the book, talks about how small business owners eventually have an 'entrepreneurial seizure.' After years of working in the technical detail of what they love—like accounting, construction, house painting, landscaping, whatever—they wake up one day and realize that they own a business!

"And for these owners to grow their businesses, they needed to get out of the technical detail and work on their business, not just *in* it. But this is counter-intuitive for most people, Brian. It's like Chinese handcuffs. The harder you pull and work to free yourself, the more stuck you get. In an attempt to grow, small business owners work more hours in the technical detail. But this is exactly the opposite of what they need to be doing."

I found the idea interesting, but irrelevant to me. I didn't know where she was headed or how this might help me and my teetering career as a CIO.

Carol continued. "Well, that happened to me, Brian. I woke up one morning and realized that I was a CIO. I had a 'CIO seizure!' I had not been able to let go of using my technical skills of architecting and building complex software applications. That had always been my value to the companies I worked for, and I felt that if I didn't use those skills, I wouldn't have anything of value to offer, and the company would get rid of me.

"While I was making sure our software was the best it could be, all of the other parts of IT were falling apart. I would pull those other managers aside and rake them over the coals, but nothing changed, nothing improved."

I was surprised by how much Carol was sharing. I rarely heard anyone admit their mistakes, especially after just meeting. I was a little intrigued.

"It wasn't until a few weeks after I finished reading that book that I saw the connection between being a small business owner and being a CIO. CIOs also wear multiple hats, and we get stuck in the technical detail. And just like small business owners, we limit the growth and profitability of our businesses by the way we function. Let me show you something, Brian."

Carol grabbed her notebook and drew this simple chart:

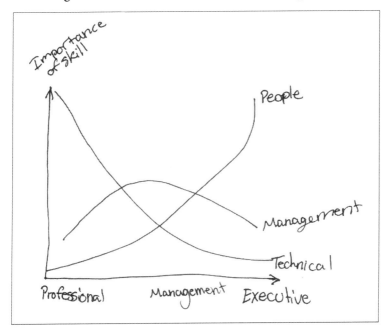

Concept from Michael M. Lombardo & Robert W. Eichinger,
© 1989 Center for Creative Leadership.

"Early in our careers, we use our technical skills the most, and our management and people skills only a little. When we move into management ranks, our use of technical skills drops while our use of people skills increases. When we move into executive ranks, as a CIO or CEO, for example, our use of technical skills is minimized and even our use of management skills drops."

"That's doesn't make sense, Carol. Why would use of management skills drop at the executive level?"

"Because at the executive level, we shouldn't be managing people, we need to be leading people. We shouldn't have to touch them on a daily basis to ensure that they're doing the right things in the right way. We should have professionals and a management team below us that we don't have to babysit. We trust them to deliver, and they trust us to stay out of the way."

I challenged her again. "If I'm not supposed to be using technical or management skills, what skills should I be using? What's left?"

"People skills, Brian. Leaders get things done through people, not so much through their management abilities and certainly not through their technical abilities. You'll get more done, infinitely more, and produce greater results for the company if you stop trying to get things done through your own efforts and work to get things done through your team."

Carol paused. "You're not buying this, are you, Brian?"

"Not yet."

"Look at what happened to me. I was ineffective as a CIO. I eventually recognized that my other teams weren't failing because they had technical problems or because they had a management problem. They were failing because they had a leadership problem. *Me!*"

I glared at Carol. I knew this discussion wasn't about her, but about me, and I let her know it. "And your point is…?"

"By operating in the technical detail, CIOs not only hurt the business but we also limit our own career possibilities. And that's where I'm afraid you are, Brian."

My glare hardened. "So your point is that I'm not a good CIO."

Carol's voice toughened. "I didn't say that, Brian. You've reached a limit, just like I reached a limit. Now something has to change."

I clenched my teeth and growled. "Are you here to do Jim's dirty work and fire me?"

Carol relaxed. "No, Brian. I'm here to help you. But you have to be open to that help. Otherwise you *will* lose your job. It's your choice."

5

The Breakfast Club

By the time Carol finished speaking, I was on the verge of bursting out of my seat and stomping out. I just wanted to leave. There was too much going on. Between my discussion with Jim, my ever-crumbling career as a CIO, and now Carol's harsh words, I was at a loss for what to think or do.

"Brian. Do you want to succeed as a CIO?"

I let the words bounce off of me at first. I heard them. I just didn't want to acknowledge them. That would mean admitting failure. But, slowly, I allowed the words to sink in, to swim around inside me as I searched for an answer.

"Sure," I finally responded, as if it was a stupid question.

Carol ignored my attitude and raised the ante. "If you continue to operate as you are now at Cantril, will you be successful?"

I didn't answer. I wasn't going to give her the satisfaction, or admit failure to someone I barely knew. I just glared at her.

Carol remained quiet as well, allowing me the grace to stew in my own anger. She was surprisingly calm. Soon the words "truth hurts" began to seep into the chaos of my mind. I settled into the discomfort of the situation while I searched for the best possible answer to her question. I took a half step forward.

"If I'm honest with you, will you tell Jim?"

"I already told you, Brian, that our conversations are confidential. I won't share anything with Jim without your permission. So no, I won't share your answer with Jim."

Carol spoke openly and honestly. She had a genuineness about her that I felt I could trust, even though we'd only known each other barely an hour.

"No," I responded. "I don't think I'm on a path for success."

"You mentioned that you're in your third role as a CIO. What happened the other two times?"

I hated reliving those events. It hurt. My throat tightened and I could sense tears beginning to form. I fought them off and swallowed hard, as if I could wash away the pain and embarrassment.

"I was fired. Both times. My first CIO role lasted a year and a half. I had promised to deliver a new CRM software solution in nine months. A year into the project we were 30 percent over budget and at least three months from implementation. But you have to understand, Carol, that we really didn't have executive support and the vendor wasn't…"

Carol threw her right hand into the air. "Brian, stop. I don't need the ugly details. You told me what I needed to know."

I shook my head in disgust, both at my previous situation and at Carol for interrupting me. She didn't even give me a chance to defend myself.

Carol pushed ahead. "What happened at the next company?"

"Well, at Nashin, I was released after two years. The president and I just didn't get along. Every time I tried to tell him about IT's needs—the fact that we were understaffed and on very old servers—he found an excuse to end the meeting. It happened often, so I just stopped trying to talk to him at all. It went downhill from there."

"When you met with the president, what did he say his needs were? What did he need from you, from IT?"

I folded my arms. "Um... I'm not sure we ever had that exact conversation." Carol glanced at my arms as if they were waving big red flags.

"You think it was my fault, don't you?" I asked.

"I'm afraid so, Brian."

I crushed my coffee cup. "Damn you," I whispered. Carol paused, but didn't respond. I forced myself to let go of the cup and relax. I was embarrassed by my action, but still angry about her accusation.

"Unfortunately, Brian, it's all too common for IT leaders to drive a wedge between the business and IT, creating a communication chasm. We in IT think IT is about technology, that our jobs are to implement and manage technical solutions and infrastructure. That's not our purpose, especially as CIOs. Our job is to help grow businesses, to help them achieve their objectives."

Carol continued to move the conversation forward, and dig deeper into my wounds. "So you were CIO at two companies for about one–and-a-half to two years each. And how long have you been with Cantril Distribution?"

I hesitated, not because I needed to do the math, but because the answer was not good. "One-and-a-half years."

Carol frowned. Maybe she just felt sorry for me. We sat quietly for a moment, reflecting on the weight of my answer.

She looked at her watch. "It's almost 10:30, Brian. Our time is about up. Would you like to get together again? It's your decision."

I still felt angry, but also embarrassed and hurt. I was drowning in a deep ocean, no longer even treading water. And there was no one to blame but myself. Carol was offering me a lifeline, ready and willing to pull me in. I simply needed to reach out and grab it.

Dragged down by the hurt and emotions, the words barely came out. "Can you help?"

"Yes, Brian, I can. But you have to be open to changing. You can't be successful as a CIO with the attitude, mindset and approach you've carried with you to this point in your career. They've clearly been ineffective for you, and, frankly, they will cause you to continue to fail. And one more important point, Brian. I can't change you—or make you change or even make you better. I can only provide the insights, strategies and guidance you need. Change is up to you."

I hesitated. I tried to think but could only react. "Okay," I said. I heard Carol's words about change, but they really didn't register. I simply knew that I didn't have a choice. "Pride comes before a fall," my mom used to say. If that's true, then maybe swallowing my pride can come before a rise. I certainly needed something good to happen in my career.

We agreed to meet at the same time each week. She didn't know how long we'd need, but suggested an initial two months. That seemed a bit much to me, but I knew I could say no at any time. I also saw the two months as a little breathing room between me and the edge of that cliff. I knew the chances of me still failing were good, so this would give me time to search for another job.

We quietly headed for the front door, and walked out into the cool morning air. A light rain was falling, which somehow helped calm the anger and hurt lingering inside me.

Carol said goodbye, then turned and headed to her car. "Why are you doing this?" I asked.

She turned, unfazed by my question. "Do you mean why am I helping you?"

"Yes. Why would you help me? You don't even know me. Doesn't your team need you? Don't you have meetings to attend, fires to fight, CFOs to argue with?"

"There is a reason I'm helping you, Brian, but I would rather you tell me why I'm doing this, and how I could possibly find the time to pull myself away from all that is happening at my office."

"I don't know, Carol, that's why I asked you." I could already tell that Carol would have a way of frustrating me.

"I know you don't see the answer *now*, Brian. What I mean is that I would rather you discover the answer for yourself. Discover why I would give up other important things to help someone I don't even know. I think you'll find the answer pretty powerful."

Carol turned again and walked away. I felt a flash of irritation. She had the answer to my question but refused to share it! But just as quickly, a stronger flash of excitement consumed me. Her words were somehow hopeful. She had something that I needed, something that I wanted.

6

That String's Too Long

I ran into Jim the next morning in the break room. I had my double tall mocha in hand but needed to toast my bagel.

"How did your conversation with Carol go?" Jim asked.

"It went well, Jim. Thank you for introducing me to her."

"You're welcome. She's a strong leader. I think you'll enjoy your time with her."

"Enjoy" wasn't quite the word I would use to describe my time with Carol, but I wasn't about to challenge him on that point.

"Jim, I need to tell you that the project to enhance our warehouse application is already a week behind. I had to move the kickoff meeting out a week in order to meet with Carol."

Jim smiled. "That's okay, Brian. Thanks for letting me know. We'll be re-visiting all of IT's projects anyway, so I'm not concerned... *yet*."

I got the point. He was joking, but serious at the same time.

"Do you have a few minutes, Brian?" Jim asked.

"Sure." What else could I say to the new CEO? I'd be late to my own staff meeting and my bagel would be cold, but a few minutes with Jim would be worth a stale bagel. As we walked to Jim's office, I asked him about the baseballs in his boxes.

"Oh, you saw those. They're from the little league teams I played on over 40 years ago. My dad coached those teams, and he always had the players sign a ball at the end of the season. He gave them to me before he passed away a few years ago."

"Wow, Jim. That's not what I was expecting you to say. They must have meant a lot to your dad if he kept them all those years."

Jim smiled, but his voice softened. "The players meant a lot to him, I know that."

As we entered his office, Jim continued sharing. "My dad never went to college, and he never earned more than about $11 or $12 an hour, but I learned more about leadership from him than from anyone else. And most of what I learned he taught me on a ball field without saying a word.

"His leadership on the field was something special. He believed in the boys he coached, regardless of their background. And he had this secret formula for building winning teams that still amazes me to this day. He knew that winning didn't come from talent alone."

Jim walked over to his bookshelf and picked up one of the plastic cases. He pulled the ball out and motioned me over. "Come here. Look at this one." He handed me the ball. "It's signed by Jimmie Hendrix."

It had been years since I'd held a baseball. I cupped it in the palm of my left hand, and then rolled it over to my right, gripping it tight with my fingers. Its weight was heavier than I remembered. It was dry and rough. The smell of musty leather leaped to my nose. I remembered that smell. On second thought, maybe heaven should have the aroma of old baseballs. The red stitching was still tight, and while the blue ink of the signatures had faded, I could still read most of the players' names.

And there it was. "You weren't kidding! I can still read it, 'Jimmie Hendrix!'"

Jim laughed. "Okay, it's not *that* Jimmie Hendrix. There were only a couple black families in our small Ohio town. And only one of their kids played in the league that year, and he was on my dad's team. Jimmie thought he was Hank Aaron. And my dad treated him like he was."

I smiled and handed him the ball. He carefully set it back in its place and grabbed another one. "This one's special too," Jim said. "Do you see this name? Danny Walker. He was a hard-throwing left-hander."

I nodded as I held the ball and read the faded signature.

"Danny grew up in a tough family environment, and though we didn't have much money, they were worse off. He came to a game one day saying he'd lost his glove and didn't have another one. He was sad and embarrassed. He knew that his family wouldn't be able to afford a new one. At the next game a few days later, Danny showed up with the biggest smile on his face as he walked into the dugout. 'Hey guys, guess what? I got a new glove! I walked out my door to head to the game and this glove was sitting on my front porch!' We asked who it was from, but he didn't know. He said there was only a note that said the glove was for him."

Mary, Jim's assistant, stuck her head in the office and reminded him that he had a sales call in 10 minutes. He asked her to get the conference call going and that he'd be there on time.

Jim pulled the conversation back to business. "How did your

meeting with Carol go?"

I didn't feel comfortable opening up to Jim just yet. "I enjoyed meeting her. We agreed to meet weekly for a month or two."

"Good to hear, Brian. To be very clear, I need to see rapid progress from you. Your initial moves will tell me if you're focused on the right things to position IT for greater business impact."

The pressure of Jim's words and expectations landed heavily on my shoulders… a 50-pound weight on each. My heart began to race. "You have my word that I will work as fast as I can. I don't want to make mistakes though, so…"

"You will make mistakes, Brian. I'm not asking for perfection. I am asking for action though. And that action needs to take IT in the right direction. You won't get in trouble for making mistakes, but you will see me in your office if you're not making continual progress."

I nodded my head in understanding. It was all I could manage. He sensed that I was still unsure of what he was asking for.

"Brian, I worked with a sales guy many years ago who didn't understand the value of marketing. He just couldn't see how it would help his direct sales efforts. After trying my best to explain it over several conversations he finally said, 'Stop!' He lifted his hands in the air and held them about two feet apart like this, pinching each hand's index finger and thumb together. Then he looked me dead in the eyes and said, 'Jim, that string's just too long!'

"What he meant was that he didn't see how marketing our services would help him with his direct sales efforts. He couldn't follow that line of thinking from one end of the string to the other. He got lost along the way."

I laughed. "Typical salesman."

Jim didn't laugh. "I know what you mean, Brian. But similar to salespeople, few technologists get the connection between IT and business performance. That string is just too long for them too! Typical CIO, don't you think, Brian?"

I was stunned. I didn't expect such a cutting comment from Jim. I know I should have laughed, but he was right. I *was* just like that sales guy. He failed to see an important connection critical to his effectiveness, and I was no different.

"Brian, leaders see connections and leverage them. I need you to see the connection between technology and the business, to be able to follow that string from IT efficiency and effectiveness to greater business results."

I looked at Jim. I could tell this was a pivotal moment. This is where he would learn whether he could rely on me or not. He sat on his desk and crossed his arms, as if he had just rested his case after a long closing argument, and was waiting for my response.

"I need to be honest, Jim. I don't see the connection. But I want this. I want to be a successful CIO, and I want to help you grow this company." I knew it wasn't the best answer, but it was honest.

"Okay, Brian. It's not what I was looking for, but I appreciate your sincerity. It's a trait I look for when I build leadership teams. It's one of the things I learned from my dad. You can't build winning teams with talent alone. The team needs to have heart as well. My dad just had a knack for knowing who had talent *and* heart. He never told me this, but I would bet every one of these baseballs that he would have chosen a player with all heart and little talent, over a player with all talent and little heart."

I was looking Jim straight in the eyes. I wanted him to know that I was listening and meant what I said about wanting to help him grow Cantril Distribution.

Jim continued. "I think you've got heart, Brian. What I mean by that is you have a strong desire to make a difference, and you have a humble nature about you. You were willing to admit that the warehouse project was already a week behind, and now you've just admitted that you don't really get the connection between IT and business results—even though you know that it's critically important to me."

"I know. I thought it was the right thing to do," I said.

"Brian, your strong technical abilities got you to this point in your career, but they're going to be your downfall if you don't change. You need a different mindset. You need to stop managing IT and start leading IT."

"That's exactly what Carol told me."

Jim laughed. "Brian, I believe in you, and I think my dad would have picked you for his team. But now you need to put your technical talent aside, lead with your heart and deliver results. Or you won't be here much longer."

I realized that I was still holding one of his baseballs. I looked at Jim and carefully tossed the ball back to him. The weight of the ball was gone from my hands, but not the weight of his expectations. I fully appreciated the opportunity he was offering, and I certainly believed in myself, but I still wondered if I could do this. I saw clearly now that Carol would be the key to my success.

Mary stuck her head back in the office. "Everyone's ready, Jim."

Jim abruptly left and headed to his meeting. I walked over to the window and stared out at the snow-capped Cascades in the distance. I was now anxious to meet with Carol. Hell, desperate would be more accurate. But could she help fast enough to save me?

7

Work Yourself Out of a Job

"It's still not done? Crap. Okay, I'll take care of it myself. I apologize for the delay, Rich. Yes, I understand. Goodbye."

Carol smirked. "Trouble at work, Brian?"

"Gee, how could you tell? That was our head of operations. One of his directors had put in a help desk request to add three new users to our distribution reporting system. Simple request, right? A week later and it's still not done. He now has three people sitting there twiddling their thumbs."

"Why are you taking on this responsibility?"

"Because it has to get done!"

"I understand that it has to get done, but why do *you* have to do it?"

"Because if I don't, it's clearly *not* going to get done! Carol, the pressure is on me to perform. If IT fails, I fail."

"I agree with you, Brian. Is IT failing now?"

I clenched my teeth. "What do you think?"

"Okay, so why is it failing?"

Here we go again with her damn annoying line of questioning.

"Because the help desk is actually supposed to help people but doesn't, my business analysts don't have good people skills and frustrate the business, the developers work on whatever they want to work on and we haven't delivered a project on time since I arrived. There, how's that for starters?"

"So your response to a failing IT organization is to be the help desk, serve as the business analyst, prioritize work for the developers and manage all projects?"

"Damn it, Carol. Why are you always pushing my buttons? Is this entertaining to you?"

She didn't answer. I looked away from her. After a few seconds of uncomfortable silence, Carol leaned toward me, pounded her finger into the table and spoke in a tough tone I had not experienced from her.

"Brian, you're the one who agreed to this help. If you can't handle me jumping into your shorts and kicking your butt a little, how the hell will you ever become a real CIO? How could you ever handle a challenge from subordinates and still move them and the organization forward? Or take a verbal beating from the CEO and walk out stronger and better prepared to make a difference?"

I was burning hotter now. I was ready to quit, just walk out. I pushed my chair back and prepared to leave. I would just have to dig in and work harder to keep my job. It would mean longer hours, but it seemed the only way now.

"Brian, if you quit now, you'll be back in the same boat with Jim. And when that boat sinks, you'll be back in the exact same boat with your next company. And then the next company and the next. How long do you plan on bailing?"

I was raging inside, but I knew she had me cornered. We must have sat there for five minutes without talking. I had forgotten we were in a coffee shop. It was just me against her, I thought. Until I realized that Carol was on my side. It was just me against me. "Damn it," I said out loud, shaking my head, not wanting to surrender.

I needed to face reality. I needed to admit the truth. I sucked in a deep breath and pushed it out slowly. Okay, I admitted to myself, the truth is, I'm not a good CIO. I want to be, but I'm not. I do need help.

Carol again broke the silence. "Do you want to leave?"

I stared into my coffee, tightly gripping the cup with both hands. The heat was almost more than I could bear, but I couldn't let go. I was searching for an answer, but wishing for an escape.

I couldn't believe I was ready to give in. I was so angry at Carol and here I was surrendering to her. The words were the hardest I'd ever uttered. I shook my head again. "No," I said. "I want this too bad. How do I get there?"

"I think you just started, Brian. The first step is humbling yourself, admitting the truth of the situation."

My anger began to subside, but I still couldn't look at her.

"Brian, you're not alone. I've been in that same boat and had to admit that I was failing as a CIO too. My problem wasn't that I tried to control and play every position in IT like you, though. Instead, I dived deep into one area and neglected the rest of IT. When those other areas started failing, I blamed them. I frustrated my team and the business because I was too busy in the technical detail to recognize what was happening around me. I didn't understand that everyone was waiting for me to fix the situation."

"Okay, so you fixed your situation. How do I fix mine?"

"Even though our situations are different, the solution is the same. In fact, it's the same even if we were CEOs, CFOs or in any leadership position. I worked myself out of a job and you need to do the same."

"What? Work myself *out* of a job? Are you freaking nuts?! I'm trying to keep my job!"

The words came out hot and angry, but Carol just laughed. "I know, Brian. It's not what you think. When I was in the detail of being a technical architect as you're now in the detail of playing every role in IT, we prevent ourselves from being CIO. When we're doing, we're not leading. Leaders get things done through people, not through their own brute force.

"You're like the conductor of a professional orchestra who's frustrated at how his pianist is playing. So you stop directing, jump down from your podium, shove the musician off his bench and start banging on the keys yourself. Then you hear the lead saxophonist miss a note so you rush over, push her out of the way and start blowing. All the while, the music starts sounding like a middle-school rehearsal instead of a world-class symphony, the actual purpose of the orchestra!"

I was still mad at her, but Carol's image forced me to break a smile.

Carol softened her tone. "Brian, your job is to direct the music, not play the notes. The only way to become a leader is to first understand the value of your value, your purpose. You're a very expensive help desk technician, business analyst or project manager. I was a very expensive application architect. And not only is it costly to the company when we play these other roles, but they fail to realize the full benefit they're expecting from their investment in us—leveraging technology for revenue growth, increased profitability and business advantage."

"But some of my musicians aren't that good. I have to do something!"

"Yes, you do have to do something, Brian. You have to stop working *in* IT and start working *on* IT. You need to work yourself out of a technical job and into a leadership job."

"Okay, Carol, I'll play along. How do I stop working in IT? And even if I'm able to do that, won't IT fall apart anyway, and even faster without my extra efforts?"

"No, Brian, IT won't fall apart. One single action on your part will help you to stop working *in* IT and it will also be your first step toward working *on* IT."

"And that action is…"

"And that action is building an IT team that you trust to execute without you, one you don't have to micro-manage or step in and save every time they mess up. You'll be able to stop working in IT when you get the right people in the right roles doing the right things the right way. And when you're focused on what I call *right leadership*, you're finally working *on* IT.

I stared at Carol, wrestling with my next question. "So I should start firing people, that's your solution?"

"You might have to replace some people, yes. You've said yourself that you can't trust some of your team members. If you can't trust them, and they're causing you and IT to fail, why are you holding onto them?"

I heard her words, but I didn't respond.

"Your job as a CIO, as a leader, Brian, is to continually build and grow a team that is effective for the business—and that works without you. That's what you need to spend most of your time doing. You need to build a management team that is capable of running IT without your daily touches or guidance."

"So, I have three direct reports. Kim Jennings runs our Project Management Office, Brett Gildersleeve runs software development and Patrick Brown manages the infrastructure group. Kim is my best manager. Her team likes her and I don't have to worry about her too much. Brett manages his group well, but we do have quality issues with our software. Patrick is a disaster. He's been here for about 12 years and no one touches him. Even I just work around him."

"Let me see if I have this right. Kim is your best manager over the PMO but you haven't delivered a project on time in nearly two years, Brett manages his group well but you have quality issues, and Patrick is a disaster. Did I miss anything?"

She made her point, and with only a hint of sarcasm. "So what do I do?" I asked.

"I want to you to go back and have a conversation with each of them. Find out if you're allowing them to be successful, or if you or something else is impeding their effectiveness. Find out what's working and what's not working for them. Ask them what their vision is for their team, for IT, for the business. Ask them if they have the right people, the right structure, the right tools to be successful."

I glared at her. She was asking me to do something that I wasn't comfortable doing. The conversation with Patrick would be especially difficult.

"I don't have a choice, do I?"

"You need to do this, Brian. I know it might be difficult for you, but you have to take a first step. Let them do most of the talking. I simply want you to ask questions. Don't judge, don't disagree, don't challenge. You owe them this. You owe them the opportunity to openly share. Just ask questions and listen. This will help you determine if you have the right management team or not."

"Okay, I'll work on it."

"I don't want you to just work on it, Brian. You need to do this. Do you think you can have this conversation with each of them before we meet next week?"

I looked down at the table and slowly shook my head back and forth. I didn't want to have these conversations. Ever. But I needed to show Carol I was serious about saving my job. I reluctantly agreed.

"I think you'll find it a rewarding experience," Carol said.

I wasn't convinced. "We'll see," I responded.

She smiled, seemingly confident that she would be proven right.

"Before we go, Brian, I want to schedule a time to begin sharing with you the three core traits of true leaders. Are you available for lunch on Monday?"

"I think so. But can't we have this conversation during our regularly scheduled time?"

"We can, but we won't get to the traits for a couple weeks. Besides, I think it would be good to have a change of venue, maybe take some of the edge off of our intense discussions."

I wasn't too keen on meeting twice in one week, but I still needed her on my side. And I was a little intrigued about the three traits. "Okay, lunch sounds good," I said.

We settled on a restaurant and time and headed outside.

I drove slowly back to the office. I needed to let the intensity of that discussion settle down. As it did, I felt further away than ever from saving my job. Not the feeling I was hoping for or even expecting. I just didn't see how it was going to happen, let alone happen fast. I needed a miracle.

8

Team Dynamics

"What were you and Brian talking about, Lisa?"

"Oh, good morning, Brett," Lisa responded. "Brian just asked me to set up one-on-one meetings with you, Kim and Patrick."

"We never have one-on-ones. What's this about?"

"He didn't say."

Patrick interrupted Lisa and Brett. "What's going on, guys?"

"Brian just asked Lisa to set up a time for you, me and Kim to meet with him one-on-one."

Patrick shrugged his shoulders. "So?"

"Doesn't that concern you at all? We get a new CEO, Brian starts having offsite meetings and now one-on-ones with us?"

"Doesn't worry me at all. I've been here a heckuva lot longer than Brian, and I've survived every other CEO and CIO change. Brian doesn't have a clue. He's so buried in the details he doesn't have time to manage anybody or understand what we do. I'm headed out to lunch. Ciao, guys!"

Lisa consoled Brett, ever worried about everything. "I'm sure it will be okay, Brett. Now if Jan in HR starts hovering around, then you can get your résumé ready!"

"Very funny, Lisa. I know you're just trying to comfort me, but I've seen this before and know how it unfolds."

"All I can tell you, Brett, is to be open to hearing whatever Brian has to say to you. And I know he appreciates honesty."

"Thanks Lisa. You're pretty wise for an executive admin."

Lisa just smiled. "Okay, since you're standing here, I scheduled you first. 1:30 this afternoon."

"What?! 1:30? That's only a couple hours away. I'm not ready!"

"He wants these meetings done by Friday, Brett. Besides, he seemed pretty serious, so I'm trying to get them scheduled as soon as possible."

"Cripes. I'm gonna go prepare."

Lisa laughed. "Prepare for what, Brett? You don't even know what he wants to talk about?"

"Yeah, prepare for what?" Kim chimed in as she walked up to Lisa's desk.

"Kim, Brian wants to meet with you, Patrick and me one-on-one."

"So?"

"Cripes, doesn't anybody have a brain? Don't you see what's going on here?"

"Brett, I see what's not going on and that's work. I'm looking forward to meeting with Brian. When's my meeting, Lisa?"

"Tomorrow morning at 8:30."

"Great. When's yours, Brett?"

"1:30 today!"

"Wow, you'd better go prepare!" Kim said. She and Lisa laughed as Brett grew more worried.

"Very funny, Kim," he said. "Hey Lisa, when is Patrick meeting with Brian?"

"It will have to be tomorrow afternoon at 2:00."

"Oh man, could you switch Patrick and me please? I'm not ready for this, and he won't care."

"It'll cost you a latté, Brett."

"Fine. Done. Thank you, Lisa."

"No problem. I'll do anything for a latté!"

"And I'll do anything to delay meeting with Brian. This has 'bad' written all over it."

9

The Conversations

"Hi Patrick. Come on in. Sorry for the short notice of the meeting."

"No problem, Brian. Will this take long? I've got a lot to do today."

"Not long, Patrick. How are things going?"

"Um, okay. They would be better if I could get back to work."

"What's working well in your group, Patrick?"

"Huh? What's working well in my group? Why would you ask a question like that?"

"Patrick, we have a new CEO. He expects more out of IT. He wants us to help drive greater business growth and profitability, even help innovate."

"That doesn't make sense. Maybe the development team can help with that stuff but my infrastructure team is too far removed from the business to make a difference. But I would love to get some R&D dollars. Can you make that happen?"

I'm an idiot for starting with Patrick. How do I get through this?

"Help me out here, Patrick. Give me one thing that's going well in your group."

"My team loves me. That's a good thing, isn't it?"

"That's not quite what I'm looking for, Patrick. Let's skip that question. What's not working well in your group?"

"What kind of question is that? What's going on, Brian? Just be straight with me."

"Okay, Patrick. Like I said before, Jim doesn't feel that IT is good enough. He wants more out of us."

"That's your problem, Brian, not mine. My group is doing well. Application uptime is high, customer websites are stable, our phone system is state-of-the-art now, and you know I could go on."

"Why did I have to step in and get application access set up for Rich's three new-hires?"

"We're not perfect, Brian. That was a blip. I appreciate your help with that one, though. Thanks."

I always had a short fuse with Patrick, and he just burned through it in record time.

"Damnit, Patrick! I get two or three calls a week asking for my help because you or your team didn't respond or follow through on something."

"Geez, Brian. You don't need to get mad. If you're feeling the pressure, don't kick your team!"

I gritted my teeth and growled. I somehow found my composure to continue the conversation.

"C'mon, Patrick, you don't feel there's room for improvement in your group? Your team's results? Your own management or leadership skills?"

"Are you questioning my abilities, Brian? I think you're going a little too far with your critique here."

"Patrick, I'm simply asking if you have room for improvement in your group or in yourself."

"Sure, Brian, and so do you and Brett and Kim. But unless you want the things that are working to start breaking, leave me alone and let me and my team keep doing what we're doing."

I didn't respond. I needed to compose myself before I said something I'd regret.

"Okay, Patrick. I appreciate your thoughts. Thank you."

"Glad I could help, Brian. Can I go now?"

"Sure."

He got up to leave, and I started to slam my head in my hands in frustration. But one of Patrick's comments came back to me in a sort of Columbo moment.

"Oh, Patrick. You're right about one thing. This is my problem to solve. Thanks for pointing that out."

"That's what I'm here for, boss."

I leaned back in my chair, took a deep breath and exhaled. That wasn't easy, but I see what Carol means. The answer is clear. Patrick cannot be part of my organization. How can IT move forward or help the business move forward with an attitude like that? His unwillingness to learn and grow will inhibit IT's ability to help the business grow.

I pulled out a mental hammer and "whack!" drove a nail into his coffin. A crude image, but it was how I felt at the moment.

* * *

"Good morning, Patrick."

"Mornin', Brett."

"Hey, Patrick, how was your meeting with Brian yesterday?"

"I think the guy's got mocha poisoning in the brain. He asked me the dumbest questions. I'm still not sure what the meeting was about. But he did thank me for my help, so that was nice to hear."

"Glad to hear it went well for you, Patrick. How do you think I'll do?"

"You're toast, my good man, toast."

"Cripes…"

* * *

"Good morning, Brian!"

"Hi Kim. Come on in. Let's sit over at the table, if you don't mind. This will be a casual conversation."

"Works for me."

"So tell me, Kim, how are things going? How's the team?"

"Well, I have to say that things aren't going well, Brian. I would really like to make some changes. I just can't do this anymore."

"Do what?"

"I've been the PMO Director here for three years, with you for the last year-and-a-half. We're still not running the PMO or projects like they should be run. But my hands are tied."

"What's not working; what would you change?"

"I have five project managers. Two are very good, one is okay and the other two are just not cutting it. I would like to replace those two. And the business analysts in the development group spend their time on application support activities. It's difficult to get them to perform project-related work. Brett won't allow it."

"You realize, Kim, that you haven't delivered a project on time since I've been here?"

"I know that better than anyone else, Brian. There are three reasons for this. I have two ineffective project managers, very little business analyst support and no support for our project management methodology. If we can fix these three things, I will be able to deliver projects on time and within budget more consistently. I know how to build and run good PMOs, Brian. I just need your support to make it happen."

"Let's do this. Document the changes you would like to make, and we'll talk about them in more detail."

"Thanks Brian. Sorry to jump on you like this, but it's been weighing on me. I was afraid you wouldn't be open to any changes."

"This is exactly what I wanted to talk about, Kim. It looks like you're one step ahead of me. And I must admit if you had asked me for these changes even a week ago, I probably wouldn't have been open to them. Change will be part of our world going forward. I'm glad you're thinking ahead."

"I'll be honest, Brian. I've been thinking about leaving. I want to make a difference, but it didn't seem like this was the place to do it. If you're open to these changes, and I could build a truly effective PMO, I would be excited and energized again. I know that I can help IT deliver more for the business."

"That's refreshing, Kim. Thank you. Is there anything else you need from me?"

"Well, I would like to talk about business-IT alignment. There is a lot more we can be doing to improve our value to the business. I've been researching and reading about it. I also met someone at a regional PMO networking meeting who is willing to share her company's approach to business-IT alignment with us."

"I have IT steering committee meetings, Kim, isn't that enough?"

"I know you do and no, I don't think it's enough. This is more than just steering committee meetings. I'd like to share more if you're open to listening."

I felt like I needed to swallow my pride. Here is someone working for me who seems to have a better understanding than I do of how IT can impact the business. Do I push back or let her run with this? She might make me look bad, but man, is she good.

"I think this is where I'm supposed to say yes, is that right?"

"Thank you, Brian. This is exciting. I'll get my report to you in a couple days and set up a meeting to talk more about alignment."

I thanked Kim, and she flew out of my office.

That certainly went better, but I was left wondering who was leading whom.

One more to go. I checked my calendar and saw that I'd be meeting with Brett later this afternoon. This would be a coin toss. Sometimes he's good; other times he just doesn't cut it. I sighed and wondered which Brett would show up.

* * *

I heard a light tap at my door. I looked up to see Brett standing outside the doorway. Wow, did he look nervous.

"Hi Brett. Come on in."

"Uh, hi Brian. Do you mind if I eat my donut while we talk?"

Somehow I found his question funny. "No, not at all. Is that your breakfast?"

"Yeah, I never have time to eat at home. Do you want me to get one for you? I will."

"Oh, no thanks. I had some toast at home. Do you like toast, Brett?"

"Brett?"

"*Brett*, are you okay?"

"Yeah. I mean, no. I mean yeah, I'm okay, and no, I don't like toast."

I couldn't help but laugh a little at his nervousness. "So this will just be a casual conversation, okay?"

Brett pushed his donut and coffee aside. His voice quivered. "Are there going to be layoffs, Brian?"

"Layoffs? Why do you ask?"

"With everything that's going on I figured something was about to happen."

"Well, I do think there will be changes, Brett."

"Cripes," Brett whispered under his breath.

"I'm being honest with each of you, Brett. Jim is not pleased with IT. The burden is on me to transform this team into an organization that can help drive greater business growth and profitability."

"Huh? Really? How do you do that?"

"I have some ideas, and Kim definitely has ideas. I'll need you and Patrick to help lead this as well."

Brett's voice cracked again. "Uh, okay."

"So what's going well with your team, Brett?"

"Um, well, I have a very talented group of developers, as you know. They're all the best we can find. Our monthly software releases have gone well over the last few months. My business analysts are also the best in town. I wouldn't trade them for anything."

"Okay, what's not working well with your team?"

Brett looked off into the corner of my office, staring at nothing. "Um, well… well, we could be doing better with our projects, but Kim's project managers don't have a good handle on them. They don't give my developers much notice about which projects they'll work on and when. And she wants my business analysts to spend more time on projects, but there's too many other things for them to focus on."

"And how are you doing, Brett?"

"I think I'm doing good. But I guess you get to decide that, right?"

"I suppose. How do you think your team could help the business grow and be more profitable?"

"Umm, well… hmmm. Umm… I'm not sure."

I could see the fear in his eyes. He's a good guy, but this was painful to watch.

"Brett, you've done a good job up to this point, especially considering the fact that I haven't challenged you. Nor have I done a good job setting proper expectations or holding you accountable. But that has to change, starting now."

"I understand, Brian. I'll do whatever I can for you."

"I need you to think more critically about your team, what's working well and what isn't. For example, it's just not believable when you say you have the best developers and the best business analysts in town. That's just not credible, Brett."

"But they really are…"

I wasn't buying it, and he saw it in my expression. "I'm serious, Brett. I'm going to be changing things in IT. We have to become more effective for the business."

Brett looked down at the table. "I understand, Brian."

He's so dejected and lost. This must be what I look like when talking with Carol. I can empathize with him, but I can't make it easy on him. His job is on the line. *My* job is on the line.

"I need you to write up a brief assessment of your team. Include a stack ranking of your resources, any organizational structure changes you'd like to make and areas for process improvement. And include a recommendation for how you would solve the business analyst problem you and Kim are having. I'll need this by next Wednesday morning."

Brett was still looking down at the table, but he lifted his eyes enough to look at me. "Okay. I'll get on it right away, Brian."

We talked a little while longer about specific developers and business analysts. He seemed to set his personal feelings aside and focus objectively on talent, fit and results. The discussion was encouraging. He said that I could consider the report done.

Brett stood up to leave. He tried to show some confidence, but truthfully, the fear and worry he arrived with still weighed heavy on his eyes as he turned to leave.

That wasn't easy, but it wasn't difficult either. The decision about Brett's future, though, was less clear than with Kim and Patrick. I needed something to tip it one way or the other.

10

Trust

"Hey Lisa, I'm headed out to meet Carol for lunch."

She smiled. "You two kids have fun."

I laughed. I could always count on Lisa to lift my spirits.

I pulled out of the parking lot and onto the freeway for the 20-minute drive. I was feeling more confident after getting through the meetings with my team. On the way, I worked on my attitude. I didn't want to get angry with Carol again. I needed to have that wall down. So I just kept repeating, "You're there to listen and learn… you're there to listen and learn…"

Carol was already there sipping an iced tea when I arrived. We ordered our meal, and she jumped into the conversation right where we left off a few days ago. "We actually spent a lot of time during our last discussion on the first core leadership trait—trust."

"I don't think my team trusts me, and I doubt that Jim does either," I said.

"People will trust you when they see you as credible and believable. When you consistently do what you say you'll do and when you fight for what's right. People want to work for leaders that they trust and believe in. When their heads are down and they're busting their butts for you, they want to know that what they're working on is of value. And they're trusting in you to make sure that's the case."

"So my team wants to trust in me to make sure their efforts are of value to the business?"

"Yes, and it's your responsibility to build that trust in yourself. But you also need to trust in them. And similarly, you need to build Jim's trust in you, and you need to trust in Jim. Notice that you're the linchpin to these trusted relationships. Unfortunately, you're in a particularly tough spot because building trust can take time, and that's not a luxury you have."

I understood. "Jim told me the same thing. So can I do this, build this trust in the limited time I have?"

"Yes, you can Brian. But don't focus on the trust. You can't take it from people, you can only earn it. And again, you earn it by delivering on your promises, making sure that your actions and results consistently match your words. That's the level of integrity that people look for in their leaders."

I understood integrity and trust, but I wasn't convinced I had the time. "I'm sorry, Carol, you're telling me that building trust takes time, but that I have to do it fast. That seems like contradictory advice."

"Brian, it's the small things that build trust, especially early in relationships or situations. Remember your first CIO experience? You promised to deliver a CRM solution in nine months. How did that work out for you?"

I sighed and shook my head. "It didn't."

"You basically told that CEO you were willing to take one

pitch… just one. And you promised to hit a home run with that one pitch. You gambled everything on one swing of the bat! You failed, and he quickly learned not to trust you."

I again felt the pain of that failure, as well as the deeper pain of my current situation. "And now I'm not taking any risks, big or small, and that's not working either. Okay, so what are some small steps I can take to build trust?"

"You've already started, Brian. The commitment to talk with your directors and critically assess them is an important first step. You'll definitely need to build on that step, though, with additional actions, both inside IT and outside of IT."

"Like what?"

"If you release any directors, you'll need to replace them with stronger resources. If you don't, you'll lose credibility. You'll need to communicate new expectations to all of IT and hold the team accountable for those new expectations. You may also need to assess current projects in terms of their chances of success or value to the business, canceling the ones that don't meet standards. Outside of IT, you'll need to build stronger executive relationships, and better understand corporate performance metrics."

I wrote down the ideas that Carol shared and read back over them. "Is this what you did?"

She pointed at the list. "That's some of what I did to rebuild my credibility, yes. But I think the single biggest change I made was actually a mindset change. Instead of thinking I was a technologist or a technology leader, I became a business leader within IT. And that one mindset change allowed me to build the trust and credibility I needed with the executive team. This is my focus now and my actions flow from and support being a business leader within IT. With this new mindset, my relationships changed, my actions changed and my results changed."

I stared at her, trying to determine whether I believed all of this or not. "Your executives trust you now?"

"Absolutely. I certainly trusted them before, but they now also trust me because I changed IT from a dysfunctional group to one that executes. We now consistently deliver on our promises, both big and small. We're all business leaders, Brian. It's just that when I go back to my office, I help drive business growth and profitability through technology and the others do it through sales, marketing, operations, HR, finance or whatever group they lead."

I grabbed my napkin and started tearing at it. I glanced up at Carol. "And this is what Jim experienced with you—and what he's expecting of me?"

She nodded. "Yes, but if he can't get it from you, he'll get it from somebody else."

I stared at my napkin and continued ripping at it. "That's becoming very clear to me, Carol."

She looked at her watch. "We're out of time, Brian. We'll have to discuss the other two leadership traits in our future meetings."

Somehow, she'd found the time to finish her lunch. I barely started mine. We each paid for our own lunch, and then walked quietly out the door.

"Is this helping, or making things worse for you, Brian?"

"I'd like to say it's helping, but I think it's too early to tell. I understand a lot of this intellectually, and I want to make this transition that you talk about, but it's still unclear if I can do it or not. I guess I don't feel much different yet."

"Take the actions we discussed, and I think you'll surprise yourself, Brian. When you start taking steps in the right direction and continue taking those steps, you'll get there, I promise. You have to believe in yourself and the process of learning and growth."

I heard her words but they didn't register. I was still trying to feel my way through that mental shift from technologist to business leader. I could tell this was going to take time to work through. It actually felt like I was turning my back on IT. But maybe that means I had my back turned on the business all along.

We said goodbye and headed our separate ways.

I sat in my car for a good ten minutes before I put the key in the ignition, letting my recent discussions with Carol, Jim and my team all swirl around in my mind. I searched for some miraculous epiphany that would catapult me out of this mess. I was simply frustrated. I finally turned the ignition and started the car.

I checked my cell phone for messages before pulling away. Twenty-three. Twenty-three freaking messages in one hour. I used to think getting that many messages meant that I was important, that I was needed. I think Carol would say that that many messages meant I was a weak leader, that I was actually an impediment, an obstacle to forward motion and productivity. I guess it's about time for me to start working myself out of this job.

11

Text Mate

"Hi Patrick. Do you know where Brett is?"

"Yeah, Brian, I think he went to a cloud computing seminar in Redmond this morning."

"His calendar shows him available."

"Geez, Brian. How long have you worked with this guy? Even when he's in his office he's not there! He's a master at avoiding work. One excuse after another."

I rolled my eyes. "Thanks for the help, Patrick…"

"No problem, boss. That's what I'm here for."

I headed back to my office. It's Wednesday morning and Brett is AWOL, and, more importantly, the report he owes me is AWOL with him. Kim delivered hers first thing Monday morning, as promised. And I actually received more than I asked for, which shouldn't have surprised me.

I slumped into my chair lamenting Brett's absence and mulling over what I would say to him. I turned and stared out my window. Hundreds of cars and trucks were zipping past the building on I-405. I envied them, heading somewhere and not a care in the world. Oh, to be headed somewhere… north or south, east or west, somewhere, anywhere other than stuck here in this situation.

"Are you talking to yourself, Brian?"

I gasped and jumped. I spun around to see Lisa standing just inside my door. "Criminy, you scared me."

Lisa laughed. "Sorry!"

I laughed a little myself. "No need to apologize. I was kind of spacing out for a moment. Do you know where Brett is?"

"Yes, he went to a data warehousing seminar downtown. He said he would be back around noon. He didn't look good, though. He asked if you were in the office yet."

"Patrick said he went to a seminar at Microsoft."

"I guess he changed his mind. Do you want me to call and ask him to come back?"

"No, that won't be necessary. I'll send him a text. Thank you, Lisa."

> B—where are u?
>
> at home
>
> home?
>
> sick
>
> but u were here earlier?!
>
> I can come bk
>
> no…where's the rpt?

The texting stopped. He wasn't answering my question. I stared at my phone's screen, anticipating his response. It felt like a life-changing moment, like the universe brought me to this point of just staring at my phone waiting for the fate of two people inextricably linked by… "Ding!"

> i didnt do it
>
> ok…hope u feel better

I set my phone down on my desk. My heart sank, but I could just sense that Brett's heart sank even more.

Damn.

"Are you sure you're okay, Brian?"

"Have you been standing there this whole time?"

"No. I left, and then I came back to tell you that Brett is at home sick."

"Yes, I know. It doesn't look like he'll be back in, either."

"Okay, I'll let his team know."

"Thanks. I'm headed out to meet with Carol. I'll be back in a couple hours."

"Okay. Do you want me to set up 30 minutes with Brett, if he's back tomorrow?"

"Yes. That's a good idea. Thanks, Lisa." I'm not sure what I'll say to him, but I don't think I can let this go. Damn, damn, damn.

I left the building for my three-minute drive down the road to Starbucks, wishing it was a three-hour drive. I needed the time to think. Everything just seemed to be crashing down around me. I was doing what Carol asked, but nothing was working. I have to tell her that I can't do this; I have to give up.

12

Vision

"Hi Brian. Brian? *Earth to Brian!!*"

Carol was leaning in front of me waving her hand. "Huh? Oh, hi Carol. How long were you standing there?"

"Just a few seconds."

She sat down, her face finally registering with me. "Sorry about that. I was thinking about one of my directors."

"Good stuff or bad stuff?"

I looked away and shook my head. "Bad."

"Want to talk about it?"

I was still shaking my head. "Not yet."

She decided not to press it and moved to our planned topic for the day. "Okay, so tell me how your discussions went with your team. Did you get to all three of them?"

"I did. None of them went as planned. My meeting with Kim went very well, but my discussion with Patrick was difficult. And my talk with Brett was good, but then he failed to do what he promised."

"Well, it's great to hear that your conversation with Kim went so well. Are you glad you did this?"

"Actually, I am. They were helpful. Thanks for forcing me to do them."

Carol smiled. "What a great experience for you, Brian. Did you learn anything in the process?"

I laughed. "That I want a hundred Kims. I could see setting them all free and trusting that they would deliver. She has so many good ideas. I've actually been suppressing her, holding her back. She was on the verge of leaving, Carol. I almost drove my most talented resource out the door! I guess I wasn't paying attention."

"It's good that you see this now, Brian. Very good work."

"And I didn't even understand how Patrick and Brett operated. Pretty pathetic, huh?"

"No, not pathetic, Brian. Well, maybe just a touch."

We both laughed out loud.

"I'm glad to see you're a little looser today, Brian."

"I really don't feel looser. I'm still somewhat lost."

"Leading a team is tough, but as you're learning, it can be fun when you have lots of Kims. That's what team building is all about—finding, hiring and developing the best talent possible. Then you get the joy of setting them free to achieve goals and visions."

Carol's words didn't ring true for me. "Joy? Are you kidding?"

"It's possible, Brian. You're simply at a crossroads. You can keep going straight, doing what you've been doing…"

"… and get the same results I've been getting… and get fired."

"Or you can turn one way and quit, or you can turn the other way and make the tough decisions to right your ship."

I stared past Carol, looking out the café window, my eyes focused on nothing. I knew the direction to take, but I still questioned whether I had the courage to take the tough actions I knew would be required down that path. I maintained my stare past Carol.

"Of course I want to be able to make the tough decisions. I want to be an effective CIO. And I want to deliver for Jim and the business and our customers." I was now able to look her in the eyes. I was asking for her help.

"Brian, do you see that Patrick and Brett are average performers at best? Patrick feels comfortable and not threatened at all by you or the company. Brett worries too much to be able to take risks and learn and grow. Yet Kim, clearly a top performer, has been frustrated and ready to leave. You're the leader, Brian. *You* created this culture. If Kim left, you would be responsible. Do you see how much IT, the business and even the customers would be losing if you allowed that to happen?

"Most companies are usually able to release the poorest performers, but they frustrate the top performers and force them out. As a result, companies are left with what I call the 'mediocre middle.' These companies actually breed mediocrity. And it's not that the people are mediocre, it's that they don't see the value in taking risks to do big things, so they fly above poor and below great. And the result is mediocre employee efforts producing mediocre company results."

I felt convicted. She wasn't talking about "companies," she was talking about me.

"Leadership is the difference, Brian. It's a very real concept that produces very real results. You can believe it or not, but you'll continue to suffer if you choose not to believe."

"Okay, so this may be weird, but I literally have this vision of a hundred Kims running around making amazing things happen. Business users are happy, executives are happy, customers are happy, and of course Jim is smiling big time."

"It's interesting that you bring up vision, Brian. That's the second core leadership trait."

I smiled, proud that I finally got *something* right. "Seriously?"

"Yep. Crafting visions and vision statements, and especially communicating them, are core to what leaders do."

"I've always viewed vision statements as touchy-feely things that made nice plaques. You know, something to read in order to pass time while standing in a corporate lobby."

Carol nodded. "For the most part Brian, you're right. But true leaders understand and leverage the power of vision."

"How?" I asked.

"Strong leaders know that their job is to move a group of people forward, in unison, toward a common goal. And they do that through a vision. But that common goal or vision has to have purpose and feel right to the team, or they won't buy into it or put much effort into helping achieve it. So it has to be something that pulls the team forward—together. Being able to craft a compelling vision, articulate it well and connect the right people to it are traits of strong leaders.

"It's easy for all of us to get lost in the details of daily business operations and mindlessly perform our work. But people really do want to know that what they do on a daily basis is important. They want to feel connected, that their daily efforts are of value… that *they* are of value."

I was listening intently. "So that old phrase 'without a vision, the people perish,' is kind of true."

"In a way, yes. Though people in business may not *physically* perish without a vision, they can and do perish emotionally and motivationally. They perish because they're not connected to a purpose bigger than themselves, so their daily efforts feel disconnected and pointless."

I reflected on her words and my tenures as a CIO. "Carol, it's becoming more and more clear to me that while I was in leadership positions, I wasn't operating like a leader. It feels more like I was just managing day to day, like it was a physical effort to *push* people forward to get work done. And I certainly never made an attempt to connect my team's daily efforts to what the business was trying to accomplish."

"Why do you think that is?" Carol asked.

"I didn't know I was supposed to. I was just going through the motions like my teams were. And I guess also because no one ever taught me. I never had a boss take me aside and teach me the things we've been talking about. I've gone to some management and leadership training classes, but I would come back and pick up right where I left off. Nothing ever stuck."

"How did it feel to talk with your directors?"

"Like I said, it was difficult, but…"

"No Brian, how did it *feel?*"

I took a moment to reflect on those discussions. I gazed into the stone fireplace off to my right. I couldn't feel the heat of the flames, but their dance gave me a calming break from the discussion, enough of an escape to pull up those feelings from a few days ago.

I drifted back to Carol and her question. "I guess it felt good and kind of energizing. It felt like I was forming or crafting something. What I was forming, I don't know, but that's what it felt like."

"Have you ever built anything with your hands?"

"Sure. I've built a few things like shelves and decks."

"Did you just start cutting and hammering?"

I laughed. "Heck no! That wouldn't be very smart. I followed plans, or at least had a picture in my head of what I wanted for smaller projects."

"Why would you want to have plans or a picture in your head?"

I shot Carol a look to let her know I thought this was a dumb conversation. We both knew the answers. But I played along. "Because I wanted to end up with the right result. If I just started cutting and hammering, I'm sure I would mess up, waste a lot of time and material and have to start over."

"So tell me why you spoke with your managers."

"Because you told me to."

"Are you going to have me hold your hand the rest of your career?"

I tried to brush that comment off but couldn't. That was a low blow. I raised my voice. "No! I don't expect you to babysit me the rest of my career." I was shocked to see that she remained so calm. I thought she would match my anger, but she didn't. She continued as if we were still having a casual conversation.

"Tell me again why you were talking with your managers."

I calmed myself down. I wanted this bad enough. "Because I want something different. I want to be a good CIO, a successful CIO. I don't want to fail again. I want to build an IT organization that works, one I can be proud of, one that delivers real business value like Jim wants and needs."

"How do you see this happening?"

Even in the face of my frustration, Carol remained unruffled. My anger dissipated as her calmness prevailed.

"We have to deliver projects on time and within budget. We also need to be more responsive to business needs. And I would also like to work on more innovative things for the business."

"Now tell me again why you were talking to your managers."

I wanted her to stop asking that stupid question, so I took a few extra seconds to think about my answer. I needed to get it right this time.

"I talked to them because I want to build an IT organization that works, one that delivers real business value. I guess I needed to know if they would be part of the solution or part of the ongoing problem."

"When you have a vision, Brian, you have something to build toward. Every move becomes deliberate in that direction. Every cut of the saw, every swing of the hammer moves you closer to achieving what you have pictured in your mind. Visions are critical. They bring people and teams together. It also gives context to the question, 'Is what I'm doing at this moment moving me closer to or further away from my vision?'"

I finally got it. "So my answer really is that I met with my team because it was a cut of the saw, a step toward building a more effective IT organization?"

"Yes, Brian. And you have many more cuts of the saw and swings of the hammer to identify and execute. But each one needs to deliberately move you closer to your vision."

"Do you have a vision for your IT organization?" I asked.

"I do, although I didn't for the first couple of years. After I had my 'CIO seizure,' I took my management team on its first strategic planning retreat. I was standing in front of the team at the end of day one talking about the importance of vision. Someone asked me what our team's vision was. I didn't have an answer. I felt like a fool. I didn't say anything for what felt like five minutes, though I'm sure it was only a few seconds. I told them I would come back with a working vision statement the next morning."

I was surprised. "You came up with the team's vision statement by yourself? Your team didn't help you?"

Carol laughed. "I'm not a fan of creating vision or mission statements by committee. I've been a part of many group strategic planning and visioning sessions and not a single one turned out well. They're frustrating and painful to sit through and the results are, well, pretty useless."

I laughed. "I can remember a particularly painful experience myself. It felt like we were all asked to work together to paint the Mona Lisa, but we all had our own brushes, paint and ideas. The result looked more like a Jackson Pollock than a da Vinci."

Carol began to laugh. When she tried to cover her mouth to keep from being too loud, she knocked over her tea, spilling what was left out onto the table.

I ran and grabbed a handful of napkins and handed them to Carol. Still laughing a bit, she slowly swirled the napkins on the table to soak up the tea.

"Okay, I'm composed now, Brian."

"I didn't know you made mistakes, Carol. You're human after all!"

"Okay, now that's not funny."

This time I laughed out loud, but kept my coffee upright and the table dry.

"Okay, back to business, Brian. Now don't get me wrong, I think it's important to get input from and collaborate with as many people as possible, but in the end, one person, the leader, is responsible for setting the vision."

"So did you come up with your team's vision that night?"

"I was worried at first that I wouldn't be able to. Okay, scared was more like it. But I eventually came up with something that everyone supported and was excited about."

Carol grabbed a fresh napkin and drew these symbols on it:

360° Δ KZN

I stared at the symbols for a few seconds trying to decipher them. No luck. I looked up at Carol and wrinkled my eyebrows.

"Okay, Brian, I get that it's a little different. It's not what I expected either. But I wanted something short, simple and memorable."

"Well, it's certainly short!"

"I got the same response from my team, but once I explained it they were excited and in full support."

"So what do the symbols mean?" I asked.

"The 360° symbol represents a leadership concept called 360°

leadership. It means that leadership isn't a position but a mind-set and a responsibility. That people can lead others below them, at their peer level and even those above them. In the context of this vision statement, it sets the expectation that we are all leaders and that we will continually develop our leadership abilities.

"The delta sign represents the difference, or the impact, we want to make in the business. And finally, KZN is our company's stock ticker symbol."

I again wrinkled my eyebrows. "I'm still not sure how this is a vision."

"So putting it all together, our vision states that by continually developing our leadership abilities, we will grow as business leaders and be better prepared to identify opportunities that will make a positive difference in our revenue and profitability, which will ultimately affect our stock price. Or more simply put, we will impact our company's stock price by continually developing our leadership abilities."

"Geez, Carol, aren't you living in a fantasy world?! You're turning IT thinking on its head!"

"Actually, Brian, I think this turns IT thinking right side up! It's been on its head way too long and the blood rushing down has created some pathetically disastrous results!"

"You really believe this, don't you?"

"I do, and so does Jim."

I nodded. "Yeah, he made that point loud and clear to me."

"I've delivered some of those disastrous results, Brian; I can admit that now. But no more. I've learned that by becoming a business leader within IT, CIOs can bring business and IT together, bridge the communication chasm and get IT focused on delivering measurable business results. And if IT organizations don't aim high, if we don't strive to make a difference in the stock price or on customer's lives, then we'd be aiming lower and *never* see opportunities to make that kind of a difference for the business."

I thought about her words. "You know, I don't think we're aiming for anything. It feels more like we're just reacting and being aimed at. I don't feel in control of IT, our results and certainly not my career."

Carol smiled knowingly. "It's good that you see this, Brian."

I looked down and spoke into the table. "This is painful to listen to, Carol, even embarrassing."

"I know that you're not where you want to be, Brian, and as the reality of your past hits you, yes, it's going to be painful to look at and acknowledge. But you have to face these facts if you truly want something different. You have to accept it and even forgive yourself if necessary in order to move forward."

I just sat there, drained.

"And you need a different foundation on which to build IT and your CIO career, which means you have to demolish your current foundation, your current way of thinking and your current way of operating."

I felt sad, but I wasn't sure why. Maybe the feeling was hopelessness.

"Brian, I'm not some Yoda or sage who was graced by the gods with these insights. I learned them the hard way. I've been where you are and experienced your pain. I've woken up face down in the CIO gutter, having to face my own failed past. But I got back up, chose a different path and rebuilt my leadership foundation."

I looked Carol square in the eyes. Something I hadn't done all morning. "I do want something different. I can't imagine being in this same boat, or CIO gutter, five years from now. I don't want to look back and think 'what if?'"

"That's a powerful vision, Brian. Not wanting to look back with regrets can drive the right actions today. And the same goes for your IT organization. The right vision for your IT team will help connect everyone and their efforts to the greater good of the company."

"So has it worked for you? Have you made a difference in your stock price?"

"Not yet, Brian, but we're very close. We have several initiatives in place that never would have come about if we didn't have this vision. But by aiming that high, we've forced ourselves to make the changes in IT that at least move us in that direction, changes that optimize IT for business benefit. If we didn't set a bigger vision we would be settling for less."

"What kind of changes did you make?"

"I'll tell you the three that have made the biggest impact. First, we implemented a business-IT alignment structure that has allowed us to consistently work on the highest-value tactical *and* strategic needs of the business."

I chuckled. "Business-IT alignment? We're aligned with the business, and I didn't need a vision statement or a new structure to make it happen."

"How do you know you're aligned with the business, Brian?"

"Because we document what they ask for, we prioritize those items and work to deliver them."

"So to you, being aligned with the business means IT is an order-taker, whatever they ask for, that's what you'll work on and deliver."

"Well sure, that's why IT exists."

"Does it? Does IT exist to do anything and everything that the business asks for?"

"Well, not everything. We don't have enough resources to do everything."

"So how do you know what to work on?"

"Our steering committee prioritizes some things…"

"And…"

"And, I suppose, we make the best guess prioritizing the rest of our work."

"By 'we,' you mean IT?"

"Well, sure," I said. "So what's your definition of business-IT alignment, Carol?"

"When IT is consistently working on the highest-value, highest-return efforts for the business. That's it. I think it's that simple."

"But don't you have an executive IT steering committee that sets strategic priorities, and isn't that enough?"

"Yes and no. Our executive IT steering committee only sets priorities for enterprise initiatives. They also resolve resource contentions on other strategic efforts, those that cross business units. For business unit strategic priorities, we let the top executives of those groups set their own priorities. We simply review them with the steering committee. Those business unit executives have a vested interest in setting their own priorities, and they never miss an alignment meeting where they can provide that guidance.

"For tactical priorities, you have to work with the right level of managers who are most affected by the daily use of business applications. They too will be glad to set priorities for you, but we set limits on the amount of tactical work we do.

"And that's the second major change we made. We decided to stop living within our means. That is, in terms of software application maintenance and small enhancements, we work with the business to limit the number of developer hours allocated to tactical support and maintenance. In the past, the more developers we added to maintenance work, the more of it we did. We decided to turn the spigot down on support and small enhancements and deliberately shift those resources to higher-value strategic and innovative efforts. IT's value to the business skyrocketed with this small tweak."

"And the third change?" I asked.

"Well, implementing the business-IT alignment model and limiting the amount of developers on tactical work were important, but they only set us up for failure."

"Huh? How were you set up for failure?"

"Because we still couldn't deliver projects on time and within budget. It's like opening a store that promises the highest-quality foods at the lowest prices, but when people arrive, the quality is no different than any other store and the prices are actually higher. They couldn't deliver on their promises... and neither could we."

"So what did you do?"

"We implemented a Project Management Office, but with a

unique approach. We did something in addition to the project management methodology and reporting and all of those typical components. We gave our PMO teeth. My PMO director is allowed to put any project on hold at any time if he sees that it's out of scope, schedule or budget and there's not a realistic plan to get it back on track. And we're still able to take a mentoring and partnering approach with our PMO as well. It's no longer a heavy-handed bureaucratic dictatorship.

"And not delivering on our promises was chipping away at IT's credibility. The only way to rebuild that credibility was to start delivering on our promises. It took us about a year to dig out of that credibility hole, but we did it. We now consistently deliver the right solutions on time and within budget."

"That's it? You implemented these three things and that allowed you to achieve your IT vision?"

Carol laughed. "Oh my goodness, no. We did many other things, and continue to, but those were the key changes we made. We also implemented a leadership development program, Lean IT, and even Agile and Scrum for certain types of projects.

"Okay, Brian, it's late. We're past our stopping time! I need to get back to the office for a meeting."

"Sure, Carol. As usual, thanks for your time. This is a lot to soak in, though."

Carol quickly packed up her things and grabbed her coat. "No problem. For your homework I want you to develop a vision statement for your IT organization. Can you handle that?"

"Can I steal yours?"

Carol laughed and shook her head. "No! You have to put your

own thoughts and effort into this exercise. Start with the end in mind, the ultimate result you want to have happen, and work backward from there."

"Okay. This could be fun."

"Good. I'm glad to hear you finally have a positive thought."

With that parting shot, Carol bolted out the door. I was left with a pile of soggy napkins, and short-lived hope.

13

Losing Customers

"Hi Jill… How's your day going?… Oh, that's good… Yes, I will need to come into the office over the weekend. I'm going to talk with my team this afternoon to get their ideas for that vision statement I told you about, and then work on it and some other things Saturday morning… Um, no, I didn't forget…. Yes, I'll be able to leave at 5:00… I love you, too. Bye."

Crap. Of course I forgot about dinner tonight. It's 3:00 p.m., only two hours left in the day and two years worth of work. I slowly sucked in a deep breath, blew it out and headed to Patrick's office.

"Hey Patrick. Do you have a few minutes?"

I walked into his office and sat on the credenza that ran below a row of windows. Patrick glanced up at me but quickly returned his focus to his monitor.

"Just a couple. It's Friday, and I want to get out early."

His response froze me. Not because I didn't expect that type of response out of Patrick; I did. But for the first time as a manager I didn't care about the hours, I didn't care that he was leaving early. Those had always been my measurements of productive employees. But now it felt different; I wanted something more. I wanted results, not endurance.

I tried to peek across the office and over his shoulder at the website he was looking at, but it was too far away for me to read. "What are you working on?"

"There's a cloud computing conference coming up in a few months that I plan to attend. It will be important that we fully understand all of the aspects of this technology so we can build our own internal cloud. Don't worry, I'll book the flights early and stay at a lower-priced casino."

What? We can't resolve help desk tickets in a timely manner, but my infrastructure director is planning a trip to Vegas!

Carol's words about being out of sync with the business steamed through my mind like a freight train. I heard the hardened steel of the wheels pound against the tracks. The weight and power of the train matched the weight and power of her words: "Visions should drive action from the top of a company down through the lowest levels of the organization. Every action within a company should tie back to the vision statement."

Patrick's activities weren't in sync with the business. I saw it. I heard it. I felt it.

And then it hit me. No vision = random activity. And no one was more random with his activity than Patrick. This was my fault, not Patrick's. I created this.

I could see that with Patrick in his role I was gambling with my career, and I didn't like the odds. I pulled out my mental hammer and—"whack!"—drove another nail into his coffin.

"What's up with you, Brian? Are you okay? You've been staring at the wall for a couple minutes. Is Jim putting pressure on you to shape up or ship out?"

I didn't quite know how to respond to him, so I kept it simple. "Yep, the pressure's on, Patrick."

Without losing focus on his website and travel planning, he tossed me a lifeline. "I got your back, coach."

Put off by his incessant cavalier attitude, I almost walked out of his office. But that wouldn't have been fair to him. I wanted to give him a chance to provide input on our vision statement. Bad idea, Brian, bad idea, I warned myself.

"One more question, Patrick."

"Make it fast, you're burnin' my Friday."

I took a breath, held it and counted. 1, 2, 3, 4, 5. After regaining my composure, I began again. "I'd like us to have a vision statement for IT, one that can connect us to the business and our customers. Any thoughts on what to include or how to phrase it?"

"What? Are you nuts? That's foofie stuff. We've got real work to do, Brian."

1, 2, 3, 4, 5, 6, 7...

"Can you humor me, Patrick? I'd like to know that my management team contributed to the team's vision statement. It's only preliminary anyway."

"How about, 'What happens in IT, stays in IT?'"

Patrick laughed. I didn't. I just burned a little more on the inside and walked out of his office.

"You said to humor you!" Patrick yelled as I walked away.

Whack!

* * *

As I passed Lisa's desk heading back to my office I could hear her laughing.

"So you think that's funny, do you?" I asked.

"I think anything Patrick does is hilarious. Don't you?"

I just shook my head in disgust at her. "Uh, no."

"Oh come on, Brian. Lighten up."

"I'd like to, but I'm trying to get input from the team for a vision statement, and I'm getting nowhere."

"And you can't have fun at the same time?"

"I thought I could, but it's not looking good."

"Have you talked with Kim yet? She'll pump some life into you."

"I'm headed there next."

"Talk with Brett, too. And I'm sure Jim would give you some great guidance."

I headed into my office. "I plan to. Thanks, Lisa."

"No problem, coach, I got your back!"

I shook my head and finally laughed. Only Lisa could get away with a comment like that.

"It's great to hear you laugh, Brian."

I glanced back down the hall and saw Kim walk into her office. I hurried to catch her.

"Hey Kim. Can I get your help with something?"

"Of course. Come on in. What's up?"

"I've mentioned Carol to you before, right?"

"Yep. And I'm jealous that you get that kind of focused coaching. We could all use some help improving our management and leadership abilities."

"Hmm. I like the idea, but we're not quite ready for that yet. We have too many other issues to address first."

"Okay, but just know that it's important to me."

"I hear you, Kim, and I won't forget."

Her eyes widened. "If you do forget, I'll remind you!"

I smiled. "So, Carol gave me an assignment to develop a vision statement for IT. I don't want to go through a formal visioning exercise, just want to develop something for my discussion with her next week. Do you have any thoughts or ideas?"

"Well, do you have a starting point, something to get me thinking?"

"It needs to represent the impact I want IT to have on the business, not so much focused on technology."

"What about 'Technology experts driving business advantage?'"

"Well, our vision statement will drive our focus and action. The first action I see this statement driving would be acquiring more technology experts. We already have that and we're still completely out of sync with the business."

Kim nodded. "I see. So, if this will drive action, I would like it to say something about alignment. What about 'Aligning technology solutions with business objectives?'"

"That's not quite right either, but this is still helpful."

We talked through a few more ideas, taking turns drafting and editing statements on her white board. After about 30 minutes, we decided to settle on some words, but no specific statement. "Customer," "value," "business," "growth" and "solutions" made the cut.

While the discussion was at least positive, nothing felt right.

I stopped by Brett's office next and had a similarly positive discussion with him. But we added only one new word: "technology," and that against my wishes. I wrapped up our talk by reminding him that we still needed to talk about the report he owed me. I told him we'd discuss it next week. Mentioning that report destroyed the excitement he showed in the vision discussion. I felt bad but… wait, no I didn't. It was the right thing to do. I can't let him off the hook, and I wasn't going to.

I walked out of Brett's office and headed toward mine. It was 4:20 p.m., and I hadn't gotten very far with my vision exercise. As I approached my office, I saw Jim standing at Lisa's desk talking with Patrick. Oh God, this can't be good.

As I joined the group, Patrick was saying goodbye to Jim and Lisa, wishing them a good weekend. He said the same to me, slapped me on the back, then turned and walked out. I just quietly shook my head.

"Hi Jim."

"Hi Brian. I just came down to personally congratulate Lisa on her three-year anniversary with us. We'll formally recognize her at our next quarterly town hall meeting, but I wanted to get in a personal thank you."

"Oh yeah. Sorry, Lisa. I knew your anniversary was this month but forgot that it was today. We should celebrate with donuts or bagels next week."

Lisa smiled. "No problem, Brian. And thank you, Jim."

Jim walked into my office and sat at my small conference table. I followed his lead. Then he stood up, closed the door and sat back down. Closing the door worried me.

He looked relaxed but didn't hesitate to initiate the conversation. "How are things going, Brian?"

"Everything is going fine. How are things on your floor, Jim?"

"I'll lie like you just did and say everything is fine with me as well."

Jim cracked a smile and then we both burst into laughter. It felt good to relieve some stress that way, but it still didn't erase the uneasiness and uncertainty consuming me.

I could tell Jim sensed my uneasiness, but instead of piling on, he shifted the discussion. He asked me about my family, where I grew up and how I thought the Mariners would do next season.

I appreciated the diversion and admired the way he so effortlessly sensed my state of mind and maneuvered the conversation to a more comfortable place for me. But even with the diversion, my mind never strayed from Carol's homework and certainly not from the sad state of my career. I was getting antsy to gather insights and guidance from Jim that might ease some of my fear and uncertainty... or confirm it.

I took the first opening in the conversation I could and pulled the discussion back to business.

"Jim, what impact do you want IT to have on the business?"

"Well, it's not changed from all of the other discussions we've had, Brian. I want our investments in systems and technology to help drive greater business growth and profitability, even help us innovate."

"Oh, trust me, your words are tattooed on my brain. But can

you be more specific; is there a more concrete way to describe the impact you want from IT?"

"Hmm. That's a fair question, Brian. I didn't think about sharing this with you, but something happened earlier this week that might help. It's something you should probably be aware of anyway."

I leaned forward in my seat. "What happened?"

"The CEO of one of our biggest and most profitable customers called earlier this week. He wanted to give me a heads-up that they will likely have to begin switching to a new logistics firm. He said our sales team did a great job negotiating better rates, but we're forcing overhead on them that they can no longer justify."

"I don't understand. How could we be forcing overhead on them? We provide transportation and warehousing services. That seems pretty basic."

"For starters, we're not providing them the information and insights they need to make real-time decisions critical to their business. They have to call our service center multiple times a day to get the information they need."

"But we have online systems and nightly data feeds for most of our customers. What more do they want?"

Damn. I know that sounded defensive. But his comments seemed directed at IT.

"I know we provide online systems and data feeds, but our competitors now offer wireless solutions and real-time informa-tion far beyond ours. In talking with the sales team, I learned that they've been trying to communicate this fact for the last few years, but no one has listened. I've seen reports as far back as four years ago that predicted a decline in contract renewals if we didn't begin providing wireless and other real-time logistics data for our customers. And I know for a fact it's been a strategic priority for the last three years, Brian, but it's never translated into IT action. This has to change. Our revenue is being impacted."

"We did have a project this year that would have provided online business intelligence for our customers, but…"

Jim interrupted me. "But what, Brian?"

"I was going to say that the vendor spent six months on it and got nowhere."

"You can't blame the vendor. That was under your watch. That effort not only cost us a lot of money that we'll never get back, but a lot of time and resource effort that we'll never get back either. And in the end, it could easily have cost us this customer."

I looked off to the side to avoid Jim's eyes. I was embarrassed and frustrated with my team and myself. Frankly, I still didn't understand why he was sticking with me.

"Brian, I know you didn't create this mess, but you haven't done anything since you've been here to fix it either. I need you to begin addressing this problem, or we'll lose more customers."

I didn't say a word. I could tell Jim had more to say.

"So now you know what haunts me about IT, Brian. Of course I worry about data breaches and system outages, but if we can't become more competitive, our downslide will continue. There's no doubt our sales are very relational, but that can only take us so far, especially in tougher economic times when companies are fighting for every bit of margin they can find.

"Brian, you know our strategic business priorities. If you can't take it from here and position IT to deliver the right solutions to solve our greatest business needs, then I will have to replace you. I don't have a choice. This isn't about you or me. This is about our shareholders and building shareholder value. That's my job, and I'm making it your job too."

My heart sank, hard and fast.

"Brian, I know you've been putting in the effort to rapidly learn and figure out what I need, but what in IT has changed since we had that first talk in my office? I know it's only been a few weeks, but I haven't seen any action. If you allow IT to keep doing what it's been doing, then the business will keep getting what it's been getting…and so will our customers. Something has to change, Brian, either IT or the CIO."

Whack! Jim pounded a nail into my own coffin. Not a good feeling from the nail side of the hammer.

I thought I had been making good progress with Carol and in my conversations with my team. I felt I was taking giant steps forward, but from Jim's perspective, I was losing ground, and that meant the business was losing ground too.

Jim's eyes pierced me. He was serious and out of patience. And I was out of time. My defenses were still up, but I knew if I came back with more excuses, it could be my last day as Cantril's CIO. I needed help, but not even Carol could help me in this moment.

I sucked in a deep breath, hoping Jim didn't notice. I slowly let it out as I tried to settle on a decision. To be successful, to even have a chance at surviving, I needed to get rid of my pride; I needed to tear down those defenses. I took another deep breath and let it out, allowing my pride to melt away. I knew I couldn't just bury my pride… it would still be there taunting me.

With each breath, I felt those walls of defense come crashing down. They had failed me to this point, and I wanted something new, even if it meant being fully exposed. It was now just Jim and me, no excuses, no pretenses. I dug deep, determined it was time for action.

I finally looked back at Jim. His intense gaze hadn't strayed from me. He was waiting for a decision.

"Jim, I need to make some changes in IT, and I need your support. I don't yet know everything that needs to happen, but I do know that I don't have the right management team. I want to release Patrick and even Brett. Patrick is an easy decision; Brett's is a little tougher. But just like you don't have time, I don't either."

Jim just sat there. I couldn't read him, so I continued with my plan.

"I also need to give Kim more responsibility and more authority. I think she is the key to getting IT in sync with the business so that we're working on the highest-value strategic initiatives. She can also rebuild our PMO and project management team so that we can deliver business solutions more consistently on time and within budget.

"I'll also need to work with your leadership team to re-evaluate all IT projects in order to make sure we're not missing any, cancel what no longer makes sense and reprioritize the project portfolio. And I'll also need everyone's support in reducing the amount of support work the development team is doing and shift some of those resources to higher-value, higher-return strategic initiatives.

"Eventually I'll need some R&D dollars if you want IT to help you win customers and stop losing them. But we're not ready for that yet."

The words just flowed out. It was all there. And it felt right.

That surprised me the most. As the words came out I knew for certain it was the right strategy. And I could tell by Jim's reaction that he was pleased with my words.

"I've been waiting for this, Brian. You have my support to make these changes, you always have. Just be sure to work with Jan in HR to get it right. And I don't need to know about every move you make. But know that at this point your words are just words. I now need your actions and results to match your words."

"I understand. Thanks for your support."

"I need to go Brian. Here's my cell number. Don't hesitate to call me if you need to talk. Know that I trust you and that you're not in this alone."

"I know. And that trust is definitely appreciated."

Jim left. I took a deep breath and let it out with a loud groan. Maybe a little too loud. Lisa came running in to ask if I was okay. I assured her that I was.

I reflected on the words Jim and I exchanged. That statement about winning customers kept coming back to me. It struck me as powerful. And in an instant, I knew that was our vision statement. I couldn't believe it. It came out of nowhere. Woo hoo! "Winning Customers." It sounded good but lacked something at the same time. I decided to run it by Kim to see what she thought.

Kim. Wow, will she be excited. But my heart sank as I thought about the impact I would have on Patrick and Brett. They will be shocked. I've released people before, but only because of clear violations of corporate policy, not because they were ineffective in their positions. And I'd been fired myself, so I know what it feels like to sit on both sides of that table. Those discussions will be tough, but necessary and important for the business.

I looked at my watch. Holy crap. It's 5:40 p.m. Jill will not be happy. I called her and explained what happened and that I'd be late. She understood. She knew what I was going through and how important that discussion with Jim was for me. I felt better after talking with her. I always did.

I hung up the phone and literally ran out of my office. I didn't even check email or voicemail. I just grabbed my coat and keys and ran.

Kim was walking out the front door of the building as I ran into the lobby.

"Hey, Kim. Do you remember all that you wrote in your report about rebuilding the PMO and implementing business-IT alignment?"

She nodded. "Sure I do."

"Make it happen. Don't wait, don't hesitate, just do it. I trust you to do the right things to ensure that IT is working on the highest-value projects for the business."

She looked surprised. "Where did this come from?"

"I'll explain Monday. I'm late for dinner with my family."

"Even people changes?" she asked.

I kept moving away from Kim toward the parking lot and my car. "Yes. Just include Jan and me in those decisions."

"Will do, Brian. Have a nice time with your family tonight."

I turned and yelled back at Kim, across a row of cars. "Winning customers."

She cupped her hand behind her ear. "What?"

"That's our vision statement. 'Winning customers.'"

She gave it about two seconds thought and yelled back. "I like it!"

I jumped in my car and pulled away.

About a mile down the road I heard a text come through. I checked my phone.

Imagine IT... Winning Customers

I smiled. Leave it to Kim to improve on something.

In that discussion with Jim it felt like I took a step beyond a point of no return. I'd verbally made a commitment to him, and now I had no choice but to have my actions and results match those words. Even though I took that critical step forward, it was a step into a thick fog. I had never been here before. I needed a new insight, something that would clear the fog and offer me greater clarity for my next steps. And although I didn't know it yet, I was about to get that clarity from an unlikely trio.

14

Food Fight

"Hi Carol. I apologize for calling on a Friday evening, but I could really use your help."

"No problem, Brian. What's up?"

"I'm headed to meet my family for dinner, so I only have a minute. The problem is that I'm not moving fast enough, and Jim communicated that to me this afternoon. Would you be open to meeting tomorrow morning for an hour or so? I could really use your guidance."

"I normally spend Saturday mornings at the coffee shop with my sisters. But if you can meet at 7:00, I could give you until 8:30 when my sisters arrive."

"Great, Carol. Thank you."

She gave me directions to the Tully's café where her and her sisters meet. I thanked her again, and we hung up.

I was about a mile from Caye's House, the restaurant where Jill and the kids were waiting. It was a 30-minute drive from the office, but it seemed like five minutes it went so fast. My mind was still racing from my discussions with Jim and Kim, but I needed to quickly shift into family mode.

I walked in the front door of the restaurant and saw my little boy standing there waiting for me. He wore a smile the size of Texas.

"Hi Big Guy! I missed you today. Where's the girls?" At six years old, Ryan was the oldest of the three kids. Bright and energetic.

"They're sitting with mommy at the table. They have pink dresses on. They look funny."

"What are you doing by the front door?"

"Mommy said I could wait here for you and show you to the table."

"Wow! You are a big guy today! Thanks, Ryan. Show me the way!"

Without hesitating, he grabbed my hand and led me straight to Amy, Marissa and their mom.

I smiled at Jill. "Hi babe."

"Hi hun. How was your day?"

"Pretty good I guess. I have a lot to share with you. In my..."

But I got the look. "Can we have family time now and talk about work later tonight?"

"Sure, Jill," I said. I didn't mind waiting. I needed the mental break from work anyway.

"Did you have to wait long for a table?" I asked.

"Yes, we did. That was so disappointing. We just sat down a few minutes before you arrived. We had reservations for 6:00 and it's almost 6:45. I've watched the manager stomp by the front door twice. He stopped once to reprimand the hostess. The next time he grabbed a tray of food from a waiter and told him that he would deliver it himself."

"Are you sure that you set the reservations for…"

"Hi! Welcome to Caye's House! I'm Cheryl, and I'll be your waitress tonight. Our specials are on a printed flyer in the middle of your menus. Here are some kid's menus and crayons. Can I get anyone started with a drink?"

We gave the waitress our drink orders. Everyone knew what they wanted to eat, so we gave her our food orders as well. Then we fell into our normal restaurant ritual. Ryan started shooting things with his finger gun, Amy was singing and Marissa started drawing colorful works of art. I wouldn't have it any other way.

Crash! Someone had dropped a tray of food by the kitchen. We were close enough to hear someone yell, "I'm delivering all food from now on! Does everyone understand that?"

My wife leaned forward and whispered to me. "That's the manager."

He whisked by and delivered a tray of food to the table next to us. "I'll be right back with the other half of your order," he said politely to the family before hurrying away.

As I looked around the restaurant, I could see the staff slowly and deliberately going about their work. They were eerily quiet but diligently working. For the next few minutes, I watched the drama unfold.

The manager was going at four times the pace of everyone else. He was walking people to their tables, delivering drinks and food and occasionally asking people how their meal was. He was polite to the customers, but rude to his staff.

I could hear the murmur of my kids' voices leaking into my

mind, but I was too focused on the manager and his activities to know what they were saying.

"Daddy, do you like chaos?" Marissa asked.

I looked at my daughter, and then shot Jill a confused look. "What?"

"She's asking if you like the restaurant, Caye's House. It comes out as 'chaos.'"

Ryan and Amy started giggling. It was very common for them to laugh at their three-year-old little sister for the way she pronounced words.

Amy proudly stepped in to correct Marissa. "It's not chaos Marissa, geez. It's Caye's House." They giggled again.

Jill jumped in to defend our baby. "Kids, you were her age once too and mispronounced words. Amy, you used to call grasshoppers 'oss-grappers' and eye lashes were 'ash-lies.'"

They didn't care. All three burst into all-out giggles listening to their mother mispronounce words on purpose.

But Marissa's mispronunciation was on target. This was chaos. And no, at this point I didn't like chaos. This manager wasn't managing, and he certainly wasn't leading. He was just doing.

And then Carol's words flooded my mind. "Be a leader, not a doer. Direct the music, don't play the notes. Let go and trust your team."

Is this what it looked like when a leader can't let go? When he can't trust his people? The chaos was compelling. Instead of being outside of the chaos and recognizing it, this guy continued to stir it up.

Oh crap. Is this what I look like at work? Is this the type of confusion I create? I was in shock. It felt like an out-of-body experience, but there it was. I saw with my own eyes the effects of poor leadership, what it looked like for a leader to not trust his people, to not be able to let go and set them free to do their jobs. I saw the chaos, the inefficiencies, the ineffectiveness that this manager single-handedly produced.

"This is chaos." I must have said those words out loud because all three kids burst into laughter in unison, all of them pointing at me.

Amy was still pointing and laughing at me. "You're talking to yourself, daddy."

Jill somehow felt the need to reaffirm Amy. "You did say that out loud, honey."

"But Marissa's right, Jill. This is chaos! This boss is horrible."

Ryan jumped into the discussion. "Is he a bad boss man, daddy?"

Before I could answer his question, Amy shocked me with hers.

"Are you a bad boss man, daddy?"

The kids giggled at her innocent question, but I wasn't laughing. Jill saw the hurt in my eyes. She quieted the kids down just as our food arrived. I ate quietly as I imagined myself creating the chaos in that restaurant. I cared and had a good heart but couldn't lead, couldn't deliver, couldn't make the tough decisions.

My daughter's question wouldn't stop echoing in my head. "Are you a bad boss man, daddy?"

I couldn't deny it. Bad may not be the right word, but ineffective? Yes. I had to admit the truth if I wanted to change, if I wanted to become a true leader. I sat there, embarrassed by that admission. But with Jim and Carol I had a second chance. I had an opportunity to redeem my past failures.

We all finished our meals. Our waitress brought our doggy bags, which, I'm sure, will sit in the fridge for a week and then get thrown away. I paid the check. Feeling a bit sorry for our waitress having to deal with this boss, I doubled her tip. It was also kind of my way of saying thank you for the leadership lesson.

As we approached the front door, the manager stopped and asked us if we enjoyed our dinner. I wanted desperately to tell him how awful his restaurant was run, that he needed to stop being a doer and start being a leader. But I didn't feel I had earned that right. And I doubt he would have heard my words.

"Everything was fine," I said. And we walked out.

I was quiet on the way home, reflecting on the events at work and the restaurant. I wondered what that manager's vision was for his restaurant. Based on his actions it would be something like "Caye's House... the place to go when you want to be seated, have your order taken and have your food delivered." Nothing more seemed to be of value to him.

He was lost in the technical detail of the restaurant and missed that people don't go to a restaurant to be seated, have their orders taken and their food brought to the table. That's just the technical detail of the restaurant business.

I was glad, though, for that experience. I could now see why I failed as a CIO. I saw my role as taking orders, buying and implementing software and hardware, and supporting that software and hardware. But that's not what the business needs or wants from IT. Business leaders want results, no different from my family wanting a relaxing and enjoyable dining experience, without the drama.

I seized tonight's images and the lessons they offered—the unmistakable distinctions between leader and manager, between leader and doer—providing me the clarity I needed about my own leadership failings. I desperately wanted that to be the end, the final class, the final lesson. But the cycle of failing and learning and growing never seems to stop, unless we choose to give up. And with help from my innocent little trio tonight, I decided not to give up. I had more to learn.

15

Humility

I walked into Tully's at 6:45 a.m. the next morning, my eyes half closed and burning; definitely not my favorite time of day. I scanned the café but didn't see Carol. I stepped in line to order my double-tall. Two of the large cushy chairs were available (who in their right mind would be here this early on a Saturday, after all!), so I grabbed one and threw my coat over the other.

I settled deep into the comfort of the chair and pulled together my thoughts from yesterday's events. I had a lot to share with Carol. But more than sharing, I needed her guidance. I knew the actions to take that would get me through the next 200 feet, but things beyond that were pretty murky. I wanted to see miles ahead, not just a few feet.

As I sat there, reflecting and gazing through blurry eyes, I noticed how well the café was running. The barista's were all active and smiling. The line was long, but moved quickly, and the patrons were happily buzzing away in conversation. Quite a different setting from Caye's House.

A familiar voice interrupted my thoughts.

"Hi Brian! I see you got here early enough to get the comfy

chairs. I appreciate that! Have you been here long?"

I shook my head. "Just a few minutes."

I grabbed my coat and threw it over the back of my chair. Carol laid her things down and took off to order her tea.

She returned a couple minutes later and plopped softly into her chair. "So, Jim said you're not moving fast enough, huh?"

"He did. And yes, I remember you telling me that he wouldn't be patient, given the situation the company was in. But Carol, I really thought I was making progress."

"You are making good progress from a learning perspective, but you haven't done anything to improve IT's results yet. So from Jim's perspective, nothing has changed, and he was probably concerned that nothing would change."

"That's the impression I got from him when we met yesterday. He told me I needed to start taking action or he would have to replace me."

I shared more of the discussion with her, including how I committed to replacing Patrick and Brett and planned to give more responsibility to Kim. I also shared with her the vision that seemed to appear from nowhere. She sipped her tea and smiled.

"This is tremendous progress, Brian. I know it's not easy, but you broke through that mental barrier that holds so many back from operating as true leaders."

"What barrier?"

"Making difficult people decisions. Leaders *have* to make tough decisions about people—release them, demote them, put them on a clear performance-improvement plan, hold them back or even promote them at the right time. Leaders who can't make these types of decisions are ineffective. They're hurting their team, the business and themselves.

"And, unfortunately, many companies have this engrained in their culture, where they're unable to make these types of difficult decisions. Underperformance reigns in these organizations—both people and financial underperformance. As CEO, Jim won't allow this to happen at Cantril, I can promise you that. It's critical that you get the right people in the right roles doing the right things the right way or you will fail, Brian."

"I know, I know. I see that now. I committed to Jim to work with HR to make these changes happen."

"Good. These types of decisions aren't easy, but I'll keep saying it, leaders get things done through people not their own efforts, and that means getting…"

"… the right people in the right roles doing the right things the right way," I said, finishing her thought.

Carol smiled over her cup of tea. "Okay, I won't beat that dead horse anymore."

I laughed. "The horse and I both thank you."

Carol rolled her eyes and cracked a small smile. "Let's get back to your talk with Jim. What else happened?"

"Not much else. I was defensive at first, but then I just felt like I needed to let go of that defensiveness, that there was a wall of pride inside me that was holding me back. Somehow I chose to allow that wall to crumble to the ground."

"Ahh, very good, Brian. That's the third core trait of true leaders."

I was surprised. "Breaking down a wall of pride?"

"Well, not that specifically, but what it takes to allow that to happen."

"And what's that?"

"Humility. So, in addition to trust and vision, humility is a core trait of true leaders. You humbled yourself, Brian. You laid yourself bare, which means you were open to hearing anything because you cared more about learning, growing and achieving greater things than hiding your mistakes and failures. You wanted this so badly that you opened yourself up to criticism. If you didn't allow that pride to crumble, you would have heard Jim's words, but you wouldn't have welcomed them, acknowledged their truth or taken any action to change. Humility is powerful."

I shook my head. "It was embarrassing, Carol. It was simply embarrassing to go through that in front of Jim."

"My guess is that Jim was waiting for this to happen. He saw your potential, but he also saw your pride at work. Good leaders recognize both. The problem is that only you can break down the pride and fill in the gap; no one can do that for you. And yes, it can be very difficult and embarrassing, but again, humility is also very, very powerful."

"Powerful how?" I asked.

"Through teaching. Humility is a powerful and infectious teacher. That's what you experienced in your talk with Jim. You opened yourself up to criticism, criticism you were willing to learn and grow from. And, when leaders allow themselves to learn and grow out in the open, they're also giving permission to others around them to make and admit mistakes, as well as learn and grow from those mistakes.

"So, Brian, humility is a powerful teacher, which also makes it a powerful team builder. When you have a team or organization not afraid to make mistakes or take risks, people are set free to give their all, knowing that their mistakes won't result in a berating or a security escort out the front door. Individual growth leads to team growth, which leads to enterprise growth."

I allowed her words to soak in, but not all of them. "Won't a team think less of their boss if he goes around admitting his mistakes all day?"

"It's not like you have to publicly announce every mistake you make! If you just go around admitting mistakes, you won't be seen as genuine."

"So, how do I know which ones to admit to? Is there a rule book I can follow?"

Carol rolled her eyes.

"Sorry," I said. "It's early, and I'm under the gun here."

"I know, Brian. Let me finish this thought, and then we need to get to the other actions you need to take in order to keep moving the ball forward for Jim."

"Okay, that would be helpful. Thanks."

"Sometimes you have to just admit your mistakes because it's the right thing to do. But there are times when you can turn your mistakes into a teaching moment. Then it's a matter of choosing the right mistakes, the right time and the right situation. You're looking for teaching moments, times when you sense your team is open and ready to learn from your mistakes. Be sensitive to those situations and take advantage of them when they come. If you put your team and the organization ahead of your own pride in those moments, you will be seen as genuine. This, in turn, will give others the permission they truly desire to give their all for you without operating in fear."

"But how will all of this help me in the situation I'm in now? Jim's lost his patience."

"You're going to replace your two directors, right?"

"I've made that commitment, yes."

"Are you open to hiring people who are smarter, more experienced, even more talented than you are?"

My eyes shot wide open at that last part. Carol caught my reaction and responded immediately.

"I'm going to have to pull out the dead horse, Brian. If leaders get things done through people, are you telling me that you want mediocre people on your team just so you don't look bad? You would rather have more Patricks and Bretts than more Kims?"

"Of course I'll take more Kims. I trust her to get things done."

"That's good to hear. While strong leaders aren't afraid of talent below them, this is a difficult step for many in leadership positions. They don't want others showing them up, but what they miss is that the right people will *lift* them up, not show them up."

We talked for a while about interviewing and hiring strategies. I committed to writing clear job descriptions that expressed not only the experience and skills I was looking for in these new directors, but also the traits and abilities I needed them to possess as well.

It was already 8:00 a.m. I wanted to tell Carol about my vision statement, but I also wanted to discuss other actions I needed to take for Jim. And I wanted to revisit the business-IT alignment discussion we'd briefly started the last time we met. So we agreed to spend the rest of our time that morning talking about my vision statement and fleshing out the critical actions I needed to take immediately. We would meet again Monday to talk about alignment.

I shared my vision statement with Carol—"Imagine IT… Winning Customers." She was excited about what I came up with, but just as excited that I didn't force it. She encouraged me to allow the vision statement to guide my hiring and my actions going forward. And if I continually checked IT's actions against it, we would be more focused on the highest-value things for the business—those that would win customers. And in the midst of that talk I realized that we couldn't just be focused on winning customers once, we needed to be winning each one, everyday.

We then brainstormed the critical actions that I would personally be responsible for and focus on in order to fix Cantril's broken IT organization:

- ☐ Work with HR to release Patrick and Brett.

- ☐ Work with HR to recruit and hire two new directors.

- ☐ Select a team member from both the development and infrastructure groups to be interim directors until the permanent directors were hired.

- ☐ Support and encourage Kim as she implemented her plan for improving business-IT alignment, project management and business analysis.

- ☐ Hire a consulting firm to assess my development and infrastructure teams in order to get an objective view of their strengths, weaknesses, risks and issues.

- ☐ Schedule a team meeting for all of IT (to be held immediately after releasing Patrick and Brett) and explain to them why I took these actions and what will happen next; the goal will be to keep them focused and engaged, as well as begin to rebuild trust.

- ☐ Make a list of all tasks I currently perform and decide which I will stop performing and which I will continue to perform.

  For those tasks that I will no longer do myself, I will take one of the following actions:

  - ☐ Stop doing all together.

  - ☐ Delegate to someone I trust.

  - ☐ Outsource it.

- ☐ Identify the critical strategic responsibilities I should be performing, the ones that would provide IT and the business the greatest return on my time.

  1. Leadership development

     - ☐ Work with HR to implement a leadership development program for all of IT.

     - ☐ Lead and execute this program.

     - ☐ Continue my work with Carol.

     - ☐ Join a local CIO peer group.

  2. IT organizational development

     - ☐ Design an IT org structure that would best serve the business and customers.

- ☐ Define the right roles and positions within each group.

- ☐ Get the right people in the right roles.

- ☐ Identify and resolve gaps in skills, ability and experience.

3. Process improvement

- ☐ Implement a continuous improvement program such as Lean IT.

4. IT Governance

- ☐ Improve the overall IT governance mechanism at Cantril, including a more effective IT Steering Committee, Portfolio Management process, IT services management, etc.

5. Relationship development

- ☐ Set aside time each week to build relationships with other Cantril execs, customers, vendors, business partners and industry councils; the goal here is to educate myself and others in order to continually pursue our IT vision of winning customers.

Carol and I agreed to review, update and prioritize this list once a month.

"You have quite a work load already, Brian, so I don't want to over-burden you. But you do need to do one more thing. You need to identify what you will communicate to your team after you release Patrick and Brett. There are some things your team will need to hear from you—how IT got to this point, why you made these decisions, why there will likely be more changes and what they can expect going forward. And, of course, you will want to talk about the impact IT needs to begin making for the business and your customers, the ultimate reason for the changes."

"So this is homework to practice humility?" I asked.

"It is in part, Brian. But there's more to it than just homework to practice humility. This is a critical communication from you to your team. You need to deliver the right message in order to keep them engaged and not scare them. You don't want to lose your good people, but you are moving into a period of time when you risk that happening."

"Do I write a speech, is that what you want?"

Carol chuckled, though I was serious. "All I'd like you to do is write talking points for what you'll say to your team. This will be the core of what you communicate when you bring them together. Take your best shot at these talking points, we can refine them together. If you can, let's meet again Wednesday afternoon to review them. That will give you a few days to develop the list."

"Were you being serious when you said that I risk losing people if I don't handle this well?" I asked.

Carol nodded. "Yes you do, Brian. If you don't communicate openly and honestly, if you don't own 100 percent of the reason IT is in this situation, even if it's not fully your fault, and if you don't come across as believable about how you're going to lead the team out of this situation, people will check out. Maybe just mentally at first, but eventually they will choose to leave physically.

"But if you effectively communicate all that you need to say, the right people will be excited and energized, and the wrong people, those who don't fit, will see the writing on the wall. And then you'll need to execute. This is where your actions need to match your words. This is the point in time when you'll truly begin rebuilding IT's credibility, as well as your own."

"I'll give the list my best shot, but I may need your help."

"Not a problem, just come prepared."

"I will. You've been a tremendous help this morning, Carol. Thank you for taking the time to meet so early on a Saturday morning—and on short notice."

She gave a tired grin. "I'd like to say it's my pleasure, but 7:00 is early even for me!"

I nodded in agreement. I gathered my things and prepared to leave, but hesitated. "Are you surprised that I'm still here, that I survived?"

"Yes, to be honest, I am. I knew that you *could* do it, but I was concerned if you would be able to do it; that is, allow yourself to make the mental transformation you're going through, which is a very, very difficult transformation for technologists."

"I appreciate your honesty, Carol. I certainly couldn't have gotten this far without your help."

"You've done the heavy lifting, Brian. I just sat here and talked. Besides, you're still in the easy part. Changing your own mindset is easier than changing people and an organization. Once you fully complete that transition from technologist to business leader, it will get tougher for you. But I won't doubt your desire anymore."

I smiled and thanked her again.

"My pleasure, Brian. *Now leave* so I can have some girl time with my sisters!"

I laughed. "Okay, bossy. Have fun."

I grabbed my coat and notebook and headed for the door. Things were starting to finally fall into place, it seemed. I wasn't as nervous about releasing Patrick and Brett as I thought I would be, at least not yet. And I was excited about rebuilding my management team, one that could run IT without me. If I had two more Kims, I'd be golden.

I stepped out into the cool November air. The sun shined brightly against a backdrop of rich blue sky. Not many November days like this in the Pacific Northwest. It was a bright new day signaling a fresh new start. Make that a fresh *false* start. I was about to learn that Carol was right. I hadn't yet completed my transformation.

16

The Transformation

"Hi Kim. It's Brian."

I decided to call Kim on my way home to get her thoughts on business-IT alignment before I met with Carol on Monday. She had been reading and educating herself on this topic for at least a year now, and she talked about it often. I wish I'd paid more attention.

"Hi Brian. What's wrong?"

"Nothing's wrong, why?"

"You never call me on the weekend, so I assumed something was wrong."

"Nope. Do you have a few minutes to chat?"

"Just a few. I'm getting ready to head out the door. What's up?"

"We…"

"Oh, Brian. I almost have my plan done for how I want to restructure my team."

"That's great, Kim. Let's get together Monday afternoon to review it. We'll need to begin executing it within the next few days."

"Really? That fast?"

"Yes. I can explain more on Monday. I'll be making other changes as well."

"Finally getting rid of Patrick?"

"What? How did you know?"

"Are you serious, Brian? Practically everyone in IT has been waiting for years for someone to deal with him."

"But his team likes him, don't they?"

"They like him because he leaves them alone. They can pretty much do anything they want and they certainly don't have to work very hard. The bar's pretty low in that group."

"I guess I really wasn't paying attention. So, yes to Patrick, but yes to Brett as well."

"Hmm. That's a tougher one, but I do understand. Don't worry, Brian, I won't say anything to anyone about this."

"Thank you for understanding that this is very confidential. Okay, can we move to my topic now?"

"Sure. Sorry for interrupting you."

"So I called to talk about business-IT alignment. You really see this as critical to our success, don't you?"

"Yes. If we do it right, we'll have a structured way for continually listening to business leaders. This will allow us to always understand and stay on top of their most important strategic priorities. Business-IT alignment will improve our communications and collaboration, keep IT working on the highest-value strategic initiatives and increase our value to the business. I just don't understand why it's been so difficult to make this a priority for us, Brian, I really don't."

"I appreciate your willingness to say that, Kim. Though I'm not so sure it's hurt us to this point, but I do think it can help us going forward."

"Oh c'mon, Brian! Look at the mistakes we've made basically ignoring business priorities. You know as well as I do that the developers work on functionality *they* want to work on. And you also know that we've pushed the business to do projects because we want to buy a new technology or new software. Remember the executive dashboard software we bought right after you started? Altogether, the software, the services and the training cost us over $350,000, and it's still just sitting there unused!"

"I do remember. That software was important to the business, but we learned too late that it wasn't the most important thing to them. They had other, more critical priorities, but we pushed this one on them anyway."

"And worse, Brian, IT viewed this as a software implementation, not a business problem to solve. If we had taken the time to first work with the executive team in order to understand the underlying business problem they wanted solved, they would have made it clear to us that there were other, more critical issues to address. If we had done this right, we never would have gotten into that mess and wasted all of that time and money."

"But Kim, your team managed that project!"

"And you blocked us from doing proper analysis around the business problem. You told me to just make it happen, Brian. You said you wanted a quick win with the CEO. So instead of delivering value, we delivered cost—cost in terms of dollars and lost resource time, neither of which we'll ever get back."

Damn. I couldn't deny her words. I had jumped into this CIO

role with the same poor mindset and approach that had failed me before. I forced IT to disrupt the business. But no more. "You're right, Kim. I can't defend or deny your words. But what I can do is lead a change in IT. Let's fix this so that we can drive business advantage instead of detract from it."

"You know you can count on me, Brian."

"I know. So on Monday, Carol will share her business-IT alignment model with me. Her approach is already working at Keoslin, so it's proven. I want to make sure that your ideas line up with hers. I don't want to reinvent the wheel here."

"I wholeheartedly support finding the best solution, even if it's not mine. Let me know what you learn from her."

"Sure thing, Kim. So I understand business-IT alignment at a basic level, but from everything I've read, it means different things to different people. And on top of this, no one really seems to be doing much of anything about it."

"All of what you said is pretty true. I've done a lot of reading and did attend that conference a few months ago. It does mean different things to different people, and there really isn't a standard model for implementation. Lots of people are talking and writing about it, but from what I can see and hear, few companies are truly making it happen."

"I don't want to just talk about business-IT alignment, I want to achieve it!" I asserted.

"I do too, Brian. I have some ideas, but they're just ideas. For me, the key to business-IT alignment is making sure that IT is always working on the highest-value things for the business, and the only way to make that happen is to talk with business leaders so we can understand those priorities."

"That's similar to what Carol said. But she wants IT working on the highest-value strategic *and* tactical efforts for the business. Her model seems to make that happen."

"I'm excited about the possibilities and will help you in any way I can. But you need to know that making these changes in personnel and implementing business-IT alignment won't solve all of our problems. We have other broken processes as well. For example, we need to fix our project management problems. I need you to give me the authority do this as well."

"It's yours, Kim. You have that authority. I'll have my own areas to focus on so you get to run with these. Just keep me informed. I don't want any surprises, and I want to make sure that Jim knows we're taking the right actions to fix IT."

"Great. Thank you, Brian. Hey, I've gotta go. Can we continue this on Monday?"

"Sure. This was very helpful for me."

"Good. Enjoy the rest of your weekend."

"You too, Kim."

* * *

A minute later I pulled into my driveway, shoved the gear into park, turned off the engine and just sat.

I reflected on my years as a CIO. I finally recognized how lost and out of my element I had been. No wonder I didn't survive in those positions and was heading toward failure in this one. I focused on the wrong things. Instead of getting the right people in the right roles doing the right things the right way, I played in the low-value tactical detail of technology, or tried to force technology on the business. I frustrated my team, the executives and, while it's tough to admit that the string does in fact reach that far, our customers.

I had been completely in over my head, lost in the detail of being in the detail. I just didn't know what I was supposed to be doing. Now that Jim and Carol have pulled me back, dragged my head out of the trees and bushes so I can take in the entire forest, I see a different IT organization, a different relationship with the business and different possibilities for business results. But the only way change will occur is with stronger leadership within IT. And that leadership can only come from me.

I want to help Jim grow this business. As a CIO, it felt strange saying this, talking about growing a business, but I guess my thinking about IT was finally right-side up.

Today had, in fact, brought a bright new start for me. I felt like I'd finally turned a corner on my CIO career. I allowed those thoughts and feelings to soak deep inside. Frozen in this transformational moment, I felt the cold outside the car slowly overtake the warmth inside.

Carol's words broke through the cold and silence. "You're still in the easy part. Changing your own mindset is easier than changing people and an organization. It will get tougher for you."

Not for me. I disagree. It felt like I had just pushed a two-ton boulder up a mountainside by myself. I was physically and emotionally exhausted. But now the boulder rested atop the mountain, primed for an effortless tumble downhill that I was about to trigger.

17

Front-end Alignment

I tossed myself into the soft, thick cushion of the chair. "This is prime real estate, Carol. How did you manage to get the nice seats by the fireplace?"

"Hi Brian! I hovered. After I picked up my tea, I saw the couple who had been sitting here begin to gather their things. I loitered around them until they left and then I pounced!"

"You didn't shove any old ladies out of these seats, did you?"

Carol just laughed.

I'd never walked into that coffee shop with more energy and confidence than I did that Monday morning. I had always felt inferior to Carol. She never said or did anything to make me feel that way; I simply allowed myself to go there. I now felt more like her equal. Not that I've accomplished what she has as a CIO, but now I know. I know what she knows. I'd faced reality, the painful truth of my situation, and met it head on. If I were a pilot, I guess I would have earned my wings.

"I know you've got a lot on your mind, Brian. But you wanted to dig deeper into business-IT alignment. It is foundational to IT success. Second to leadership, that is!"

I laughed. "Of course."

"It continues to amaze me how much business-IT alignment is not only misunderstood, but also how few IT organizations do anything about it."

"Why do you think that is?"

"I'm not sure. Maybe it's because we can't download software for it and, with the double-click of a mouse, magically align ourselves with the business."

"That would be perfect!" I said.

Carol nodded her head. "Unfortunately, there's no software that we can install to make this happen. The key is communication. We simply have to be better at communicating and collaborating with all levels of the business."

"Well, there's our problem," I added. "Technologists are better at communicating with programming languages and operating systems than with people!"

Carol laughed. "Yes, IT organizations have always had a difficult time communicating outside of IT, and, frankly, inside as well. That's why it's very common for us to decide for ourselves what's best for the business, what we should be working on."

I shook my head. "I don't think people in the business help much either. They don't come to us seeking to engage in strategic dialogue."

"So I guess that puts the business and IT at an impasse, doesn't it? We've created a communication chasm that neither side is working to bridge. But somebody has to blink. For me, Brian, it has to be IT."

I looked at Carol. I worked hard to read her thoughts. "It's a leadership thing, isn't it? That's why, for you, it has to be IT that blinks."

"Yes, it's about being a true leader. We've talked a lot about humility, so being the one to take that first step toward bridging the communication chasm is an act of humility. But there's another leadership trait at work here. Leaders also seek to listen and understand before they seek to be understood. In order to listen and understand in our situation, we first have to take the step to engage."

"So what does business-IT alignment look like in practice? Do I just walk around asking people in the business what they want?"

Carol busted out laughing. "Sorry, Brian. IT absolutely should not be order-takers, just writing down whatever anyone in the business asks for. I know you're not being serious, but this approach to managing business demand is very dangerous. You'll sign up for more requests than IT will ever be able to deliver, which means you'll be setting improper expectations and setting IT up for failure."

I sighed. "I think we're there now."

"Most IT organizations are. And because of the rapid-fire rate at which requests come in, IT organizations also usually fall into fire-fighting mode as well—where resources work on low-value, low-return efforts because they're focused on solving immediate requests. A formal business-IT alignment approach will pull IT out of both order-taking and fire-fighting modes, and increase IT's value to the business."

"Why a formal approach and what does it look like?"

"By formal, I mean that there are specific processes involved, processes that, when followed, allow collaborative sessions with business leaders to build on one another. And this is critical for an effective business-IT alignment implementation—*ongoing* communication, education and collaboration between specific business and IT resources."

Carol pulled out her notepad and started writing. "Okay. I'll describe our business-IT alignment model for you, but here are our core operating principles."

- Effective alignment increases IT's value to the business

- Pursue tactical, strategic and innovative alignment

- IT strategists are the key to strategic alignment

- Business systems analysts are the key to tactical alignment

- Innovation is free

- Maintain a current/accurate Strategic Project Portfolio

- Acknowledge contention — unlimited business demand vs limited IT resources

- Don't be order-takers

- Continually create capacity to do more

- Execute—deliver on time and within budget

- The CIO must be a business leader within IT

- The CEO must be an IT leader within the business

- Executives must view IT as an investment center, not a cost center

- Alignment is part of an overall IT governance program

- Alignment requires ongoing communication, education and collaboration

- Be consistent and persistent with alignment meetings

Carol continued describing their structure. "We break our business-IT alignment model into two parts within IT—tactical and strategic—and align each of those with the different business units within the company.

"Our business units are Sales, Marketing, Finance, HR, Manufacturing, Distribution and of course IT. We have a large enough IT organization, so we can assign a business systems analyst to each business unit. These analysts lead regular prioritization sessions for tactical systems requests with key partners in the business.

"And we have IT strategists as well. Think very senior-level business analysts. They're partnered with the heads of each business unit. This model doesn't work if the SVPs aren't involved. They have to own strategic prioritization. We assign one IT strategist to each SVP. Some of our strategists do cover multiple business units."

"So what do these strategists actually do?"

"They're the linchpins to alignment. They help build our strategic project portfolio and ensure that the only strategic efforts that get into IT for execution are the highest-value initiatives, the ones identified within each business unit. But remember, they're not order-takers, so they just don't meet with the business leaders and write down what they want. They're responsible for uncovering new opportunities, building business cases for each possible initiative and leading regular prioritization sessions with their assigned business units.

"This is where ongoing communication, education and collaboration come into play. Ongoing communication means regularly scheduled meetings. Ongoing education refers to the fact that the business continually educates IT about their needs, opportunities and problems, while IT continually educates the business about the technology available in-house and on the market.

"Ongoing collaboration simply means that everyone acknowledges that these meetings are about collaboration, that there's a give-and-take going on—not just one-way communication. It's this ongoing collaboration that, over time, produces innovation—new ideas to leverage technology that otherwise would never have been uncovered. Because innovation is simply a by-product of the effort to strategically align IT with the business, innovation, in essence, is free."

"But doesn't your IT Steering Committee prioritize all strategic business projects?" I asked.

"No. Only the top several projects from each business unit are even reviewed by the IT Steering Committee. The IT strategists each month lead their business units in re-prioritizing that unit's strategic projects, but we only prioritize the top seven or eight or so. While all strategic projects are in the portfolio, only the top several from each business unit are actually reviewed by the Steering Committee.

"Sometimes the Steering Committee will re-prioritize a project, but usually only enterprise projects. They typically leave the business units alone to set their own priorities. They will also resolve resource contention when projects across two business units are vying for the same resources at the same time.

"So that's our alignment model in a nutshell. It took us several months before the individual collaboration meetings felt comfortable and productive. If you go this route, don't get frustrated and don't give up. There is a payoff if you stick with this. You also need the right IT strategists. The best have a background as consultants, in addition to technology and business experience."

Then it clicked! I had an epiphany that tied everything together. "Oh my goodness! I can see these IT strategists being kind of like the 'boots on the ground' for helping grow the business. I see why you call them the linchpins to alignment. That means they're also the linchpins for IT helping to grow the business."

Carol grinned and nodded.

I continued with my excitement. "So I don't need to carry that burden all by myself. They can be extensions of me communicating and collaborating with the business on a daily basis."

Even Carol showed her excitement now. "Yes! That's exactly what they are, Brian! They are 'teeing-up' the highest-value business priorities, and then the rest of IT executes, delivering them on time and within budget. That's the rhythm you want to be in."

This revelation truly excited me about business-IT alignment.

"I do have two final words of caution for you," Carol said. "First, strategists cannot be resources on projects. If that happens, they'll get sucked into project work fulltime and you'll lose your ability to align with the business. And second, they cannot make project commitments; only the project teams can do that."

"Got it," I said.

"And I'll stress again that while working on the highest-value strategic initiatives is critical, being able to consistently deliver them on time and within budget is just as critical. Business-IT alignment demands success in both areas."

"Can I take these notes with me and review them with Kim? And I may need to have you or someone on your team talk with her as well as we plan our own implementation."

"Oh sure, please take them. Just know that not everything we do will be a fit for your situation. You have a smaller IT shop than we do, so, as an example, you may not be able to have both business systems analysts and IT strategists."

"Yeah, I was thinking the same thing. I don't think we could afford the full model you've implemented."

"Oh, and when we rolled out our alignment model, we said we were getting a 'front-end alignment.' The nickname gave us some fun and funny moments, and in some ways, it is like aligning the front end of a car. When you're out of alignment, you're operating less efficiently and less effectively."

"We are definitely out of alignment!"

"We were too, Brian."

"Okay, now I have a couple questions about the operating principles that you listed."

"Ask away," Carol said.

"You wrote down 'continually create capacity to do more' in

your list. What does that mean?"

"Would you agree that the more time and resources that IT invests in strategic efforts versus tactical efforts, the more value they deliver to the business?"

I nodded.

"So wouldn't it stand to reason that we should continually be searching for ways to shift resources and dollars from tactical efforts to strategic efforts?"

"Makes sense to me," I said.

"How much effort do you and your IT organization put into doing this, Brian?"

She knew the answer was none. "You're mean," I said.

Carol still wanted an answer. "So you're saying it's…"

"None. Zip. Zero. Nada. There, does that make you happy?"

"Very happy, Brian. Thank you for indulging me."

I stuck my tongue out at her. Not very professional, but it felt good.

Carol shook her head and muttered something about immaturity. "So back to our adult conversation. Creating capacity to do more simply means finding ways to reduce tactical efforts and then deliberately shifting those resources and dollars to higher-value, higher-return efforts."

"Okay, you're gonna have to give me an example."

"I'll share a simple one with you. A couple years ago we initiated a 'Drive to Zero' effort within IT. We had one software application where we didn't have good data edits. But do you think we took the time to fix the bad edits? No way. We didn't have the time to do it right. So instead, we spent about 20 hours of developer time each week writing code just to fix bad data in the database.

"Our 'Drive to Zero' campaign forced us to document all of the bad edits and knock them off one by one. That effort paid for itself in two months. Then we deliberately shifted half of that developer's time to strategic projects instead of allowing him to get fully sucked back into lower-value tactical work.

"It's not a big one, Brian, but it gives you an idea of how it's possible to create capacity to do more higher-value work for the business."

"I think that's pretty cool. But the team needs to be looking for these, right? They have to understand that this is of value to the business."

Carol stared at me and widened her eyes. "Yes, and who do you think communicates this message to the team?"

I mocked her by widening my own eyes. "Oh, I don't know. The janitor?"

Carol grimaced. "Do you have another question?"

"Yes, I have another question. I know what you mean by the CIO being a business leader within IT, but what about the CEO being an IT leader within the business? What's the point there?"

"So, I simply use that as a balance to the CIO being a business leader within IT. To fully bring IT and the business together, it takes both—business leadership within IT and IT leadership within the business. But the CEO doesn't have to become a technologist, that's not at all the point."

I laughed. "Could you imagine Jim trying to become a technologist?"

"No, I couldn't. It's not his desire nor is it his role. But he does have to hold himself and IT accountable for results. It's his responsibility to ensure that he manages the investments made in systems and technology and ensure that the company realizes a return on those investments. That's what Jim was doing by setting higher expectations for you and letting you know that if you couldn't deliver greater value for the business, he would find a CIO who could, a CIO he trusted."

I was still laughing a bit at the thought of Jim becoming a technologist. "So, without even knowing how to spell DBA, Jim is a technology leader within the business?"

Carol still didn't laugh. "I wouldn't phrase it quite that way, Brian, but I suppose you could. But as you've experienced, Jim takes this responsibility very seriously."

"Yes, he does, Carol!"

My joke about Jim somehow threw us into an awkward pause in the conversation. I thought maybe she really was upset about it.

"You won't tell Jim about me saying he can't spell DBA, will you?"

Carol hid her hands under the table. "I suppose not, Brian."

"You have your fingers crossed, don't you?"

Carol grinned, but didn't say a word.

"We're done for the day, aren't we?" I asked.

"You're definitely done!"

We both laughed as we gathered our things and headed for the door.

"No homework for me?"

"No, but we still need to review your last homework assignment—your talking points for your team. And there is one other critical topic we need to discuss before you release Patrick and Brett."

"What topic is that?"

"Change management."

"Oh, I just throw my change into my kids' piggy banks. That's the extent of my change management."

Carol grinned while covering her eyes and shaking her head. "Goodbye, Brian. Go to work, and leave the comic relief to professionals!"

"My kids don't laugh at my jokes either," I said. "Maybe it is time for me to get out of the funny business."

Carol turned and walked toward her car. "That's the best career decision you've made yet, Brian!"

Ouch.

18

Drip, Drip, Drip

"I don't recommend doing it that way, Brian," Jan said.

I had scheduled a meeting with Jan, our VP of Human Resources, to get her help with the personnel changes I wanted to make in IT. I explained that I wanted to release Patrick and Brett, and then we would likely release more once the new directors were on board and had a chance to assess their teams.

"Why not?" I asked.

"Because it's like water torture. You release two people now, and just as your team settles back into their work, you release a few more. Even if you communicate that there will be no more changes after the second round, they will still be looking over their shoulders wondering if more is coming."

"Any other drawbacks to my approach?" I asked.

"Well, think about this. After the first round, they will ask you if there will be more. You will have to say yes. It will become a fearful, stress-filled and unpredictable work environment. I just can't support this approach, Brian."

"Okay, so what do you recommend?

"I'd like to see you assess the full team and make all of the changes you want to make now, not over a series of weeks or months."

"But how could I assess Patrick and Brett's teams while they're still in their roles?"

"You don't have the insights into their teams to make the decisions on your own?"

I shook my head. "No, not enough."

"I can help you," Jan said. "Is there anyone else in IT that you trust to help? Or anyone in the business?"

"I could certainly lean on Kim."

"You don't have to tell people what your plans are, you can just ask generic questions and still get most of what you're looking for. You know there are a few people in the business who are very vocal about certain resources in IT, both the good and the bad."

"Thanks Jan. I think you, Kim and I could get through most of the team on our own. We could look at past performance reviews and get input from others when necessary."

"I think that will work, Brian."

"Can I grab Kim and begin this discussion now?"

"Sure. This is important. I can make some time for you now."

When Kim and I returned to Jan's office, she had printed off a list of all IT resources. We quickly identified the "A" and "D" players. The "B" players were fairly obvious, but we did have some disagreements. The "C" players were almost as difficult. We agreed that the Ds would be released. We struggled with a couple of the Cs, whether they should be Ds and released or retained.

We read through performance reviews, but never had to lean on other executives. That was a relief to me. There were some contentious moments though. We agreed to meet again in a couple days to confirm our decisions.

We settled on releasing 12 team members, about 10 percent of the staff. In addition to Patrick and Brett, we would release two project managers, a business analyst, two help desk resources, a network engineer, our data center manager, a help desk manager, and four developers.

We agreed that we would take this action in two weeks, on a Friday morning. Jan said she would get all of the paperwork completed in time, including the severance packages, outplacement services, legal reviews and Jim's signatures.

I shared with Jan and Kim my plan to move Steve Douglas into Patrick's role. Steve was our network engineering manager. He had a good reputation, got along well with everyone in IT and his team respected him. Steve didn't have the best relationship with Patrick, but then, that might be a good thing.

I also informed them that I selected Hyder Singh to fill Brett's role in the interim.

They supported both decisions.

We planned out the events of the release day, almost minute-by-minute. It stretched into a long afternoon and early evening, but we all felt comfortable and confident that these decisions would form the turning point that IT needed in order to reach its potential.

I thanked Jan for her time and followed Kim to her office. "We're not done yet," I said.

"Brian, I've got project reports I need to get out by this evening."

"Can we spend just 30 minutes discussing your plan?"

I saw the frustration in her face but knew she wanted to move forward with her ideas as well. "All right, I'll give you 30 minutes."

She printed off two copies of her plan. It was more encompassing than I had imagined. Instead of just a PMO, she wanted to turn her group into a broader IT strategy team. It would also encompass business-IT alignment, strategic IT planning, IT governance, portfolio management and business intelligence. That last one really surprised me.

"Why business intelligence?" I asked.

"Because we should be viewing data and information as strategic to business growth and profitability. We can keep tactical reporting in the development organization, but if we elevate BI and corporate performance reporting to a strategic group within IT, it will finally get the attention it deserves."

It was a different approach, but I couldn't argue with her point. I told her that in principle I supported it, but that this would take a couple more discussions to fully understand and agree to. We were still a few weeks away from a re-org anyway, so we had time.

Kim looked at her watch. "Okay, Brian, it's been about 30 minutes."

"I can take a hint. I'll schedule a follow-up meeting. Thank you for all of the thought and hard work you put into this. I know you've wanted to do this for a long time, but I just wasn't listening. I appreciate your persistence. This is going to be crucial to IT's turnaround."

"Thank you. And I appreciate your support and encouragement."

I left Kim's office and headed down the hall to mine. I

detoured into the kitchen and poured a cup of lukewarm, half-burnt coffee, more out of habit than thirst.

I set my cup down on my desk and slumped into my chair, completely worn out from the day. I had no desire to write my talking points for Carol, but I needed to at least get started. With a still-busy, cluttered mind, I grabbed my mouse, opened Word and started typing.

19

This Side of Change

It was Wednesday afternoon. My meeting with Carol was only 30 minutes away. I pulled up the list of talking points I had documented over the last couple of days and reviewed them one last time.

- I've made some changes in IT that I need to communicate.

- We released 14 IT resources today; these were good people with good skills, but they are not a fit for the organization going forward.

- Patrick and Brett were among those released.

- Steve will be our interim Director of Infrastructure and Hyder our interim Director of Development until new directors are hired.

- As I reflect back on my two years here, I've not led this team well, and that's hurt IT's ability to make a difference for the business; this will change starting today.

- We can and will help the business grow and innovate.

- And this isn't just a promise; the plans are in place to make this happen and you'll hear more about those plans over the next few days.

- These were not easy decisions, but they were necessary decisions.

- I can communicate that we do not have any more changes planned, so this is our team going forward.

- I know that this is a time of fear and uncertainty for some of you, but I can assure you that this is the team I want going forward; what I need from you is to stay focused on your priorities, stay focused on making a difference for the business and our customers.

- I know how it feels to lose teammates; I've been where you're sitting now.

- This will be a time of flux for a few weeks as we shift our mindsets and our actions in order to deliver greater results for the business; there will be a slight re-org coming in a week or so; it won't change what most of you do, just maybe on which team you work.

I didn't make any further changes. I felt good about what I had come up with, but I didn't have any illusions that I hit all of the right points. I was anxious to learn what Carol thought of the list. I printed off two copies and ran out the door to the coffee shop.

* * *

Carol had finished studying my list by the time I returned with my coffee. "First, Brian, I would drop the line about knowing how this feels. This may be true, but they don't care. You did this to them, not with them. You're not one of them in this situation."

"Okay, what else don't you like?"

"Brian, this really is a good list. I'm very impressed. Just because I came out with a negative comment first doesn't mean anything. That's actually the only point I would remove. I have a few to add as well."

"Sorry for the attitude. I was just pretty happy with the list."

"And you should be."

"Okay, Carol, my pen's ready, what should I add?"

"Let them know that Jim supports these changes. And ask that the team give Steve, Hyder and Kim their support during this time of transition."

I wrote down her additional points.

"I really like how you phrased your humility. Be sure to speak genuinely and from the heart, especially on this point. And finally, invite them in, Brian. If they're not sure what they're focused on is making a difference for the business, give them a way to provide that feedback to you or to their manager. Most people would be afraid to mention such a thing. Make it a safe environment for them to do this.

"And add something at the end that is at least a little inspiring. Maybe something like, 'And finally, if you want to make a difference for the business and our customers, then you're in the right place, you're in the right company and you're definitely in the right IT organization.' Tell them that you see the business cheering them on one day. And definitely mention your vision statement."

"Wow, this is good feedback. Thanks. Can I steal your words?"

"Use what you want, but remember, if what you say isn't heartfelt, your team will see through you, and you'll lose credibility. It will be hard to dig out of that hole, trust me."

"I understand. I'll make sure all of these words are mine."

"Good. I do think this is the message that will keep the team focused and together. The fear and uncertainty will still be there, but will subside as they see you follow through on your words with the appropriate actions."

"I must confess, Carol, the list came out of a few discussions with many different people. I had some pretty diverse perspectives to lean on."

"That's good to hear. But there's nothing to confess. You were open and listened to some experienced people at a critical time. That's good leadership."

"Thanks. So the list is good now?"

"It is, Brian. Let's move on to change and change management. I'll start with a joke."

I smirked and leaned forward, anxious to hear Carol tell a joke.

"Okay, there's a pirate ship that's been adrift at sea for three months. They've had to endure the same grub every day, the same rotten whiskey, the same clothes, even the same underwear and nasty smells during that time. The crew was getting restless, and the captain knew he needed to do something or there'd be a mutiny. Finally, he gets an idea that he thinks will save him. He gathers the men on the top deck to announce his idea.

"'Men, I appreciate your patience and endurance. I have an exciting announcement.' The pirates leaned closer in antici-pation. 'Men, everyone gets a change of underwear.' The men cheered. It wasn't much but it was something. They hadn't been this happy or excited in the nearly three months they'd been at sea. Anticipating a fresh pair of skivvies, they then hear the captain yell out, 'Johnson, you change with Smith. Williams, you change with Lee. Roberts, you change with Diggins…'"

I shook my head in disgust. "Oh, that's bad," I said.

A second later we busted out laughing. The café immediately got quiet except for our laughter. Everyone was staring at us, but neither of us cared. It was a fun moment with Carol, someone I'd had some pretty intense moments with over the last several weeks.

"Yes, Brian, I admit that it's pretty bad. But of course there's a point. You're going to be leading your team, and the business for that matter, through a time of change. If the change you bring about doesn't match the vision the people anticipate, you may have a mutiny on your hands."

"Okay, so how do I make sure this doesn't happen?"

"Be sure that the vision you communicate is the right vision for this time."

"I think it is, Carol. Do you doubt it?"

"It's not for me to support or doubt; you need to have the confidence that it's the right vision. Share it with Jim and get his support. I just want to stress that you are striking a vision.

Jim, your team and rest of the business will expect you to take them there. That's quite a load to carry, but that's what leaders do. They set and communicate a vision, build a team they trust and that trusts them, and lead with humility.

"So you likely won't have a mutiny on your hands with your vision. But the mutiny could begin if your actions and results don't match your vision and words at any time. That's leadership as well. You have to start the action, stay the course and push through to the end. Don't give up. Lean on Jim and me if you need guidance or a listening ear."

"I'm used to roller coaster rides, Carol. I think I'm ready for the ups and downs."

"Don't lose sight of your vision during the ups or the downs. It's easy to give up during the lows, but it's just as easy to become complacent during the up times. Leaders stay above the forest and vigilant, cheering the team on but also looking for times when adjustment is needed. And for a while, many on your team will be paralyzed by fear and uncertainty. Your job is to alleviate that fear and uncertainty as much as possible so that you don't lose people emotionally or physically."

"And how do I do that?"

"First, keep everyone busy and focused. You don't want idle hands or idle minds. Reassure them that no more changes are coming, if that is, in fact, true. Next, keep taking actions that fulfill your promises to position IT to make a greater impact on the business. As long as your actions are moving the team in the direction you communicated to them, they will continue to put more and more trust in you.

"And know that this is *your* credibility at stake, not theirs. They already know this. Eventually it will become their credibility as well, but that will be their choice after you've earned their trust and they're willing to tie their credibility to yours. You'll be on a tightrope for a while, Brian, a few weeks at least, if not a few months. You'll be up there by yourself, with both the business and IT watching you.

"It will also help you to know that some people will never make the mental or emotional shift to this change. They will fight it all the way. Most will make the shift over time, as they continue to build a new trust in you. After all, you admitted that you got it wrong the first two years, so why should they believe you now?

"And finally, some people, the smallest group, will be on board with you immediately. They will trust you out of the gate and give you 110 percent. They will be energized and excited. You need to leverage these individuals as much as possible. But they will drop their trust in you like a hot pocket straight out of the microwave the moment your actions don't match your promises. Your integrity will be shot immediately."

"And if I can't pull this off, it will be like that pirate ship was carrying red paint and crashed into a pirate ship carrying blue paint."

"Okay, I'll bite, what happened?"

"Everyone was hopelessly marooned."

She put her hands to her face and just shook her head. I knew she was laughing on the inside.

"All joking aside, Carol, I do understand the seriousness of this. I know what's at stake for me, IT, the business and our customers. I truly appreciate what you've guided me through, and for offering to continue that guidance. Thank you."

"It truly was my pleasure, Brian."

* * *

I enjoyed that afternoon with Carol. I saw a side of her that she had kept hidden. It lightened my heart, but not my burden. The next few days were tense, and intense. IT seemed more quiet than usual. The atmosphere was somber, as if everyone sensed what was coming.

20

Change Happens

The office was eerily still that morning. The conference rooms were empty and everyone was at their desk quietly working. Even Brett managed to arrive by 8:00 a.m. As I walked into Patrick's office and asked him to join me, he dropped his head and responded, "Sure." He knew already. They always do.

"He was already starting to tear up. I didn't offer him a tissue. Guys just don't acknowledge other guys crying." I thought about all of the times Patrick made me laugh, and the many more times he pissed me off. All good memories now, I suppose.

We sat in silence. I wondered what he would do next. How long he would be out of work. I thought about whether his wife would be disappointed in him or wholly supportive. But mostly I wanted to stop this right now. I wanted to go back to that Friday afternoon conversation with Jim and keep my mouth shut, just let him fire me. I didn't want to be here doing this.

But Jan walked into my office, and I knew there was no turning back. Everything fell into place after that.

"Hi Patrick. Hi Brian."

Patrick looked down. His voice carried a sadness he had never exposed before. "Hi Jan," he said.

I was quiet, yet resolute. "Hi Jan."

I began. "Patrick, I've decided to make some changes in IT. I'm sorry, but your employment with Cantril Distribution is ending today. I've discussed this change with Jim, and he is supportive of my decision."

I stopped there. I allowed him time to hear those words. He didn't move. He just stared at the table, his arms straight, his hands tightly grasping the chair. He was emotionless, except for the single tear falling down his left cheek.

He responded as if he'd been here before. His voice was somber. "Okay. I understand. I don't have any questions."

I had prepared more to say, but more would have been wrong. It was that simple, that quick.

"I'll let Jan take over from here. She'll review your severance package and the support Cantril will provide you."

He didn't say a word. He looked at Jan, letting her know he was ready to listen.

Jan talked for about 10 minutes, compassionately describing each document within his package, caringly answering each of his questions.

"You're welcome to pack up your things today or schedule a time over the weekend to come back in. I can meet you here, and you can gather your things in private."

Patrick said he'd prefer to pick up his things now and asked if he could say goodbye to a few people. Jan agreed. She collected his badge and cell phone. Then Patrick stood up and left my office for the last time.

I walked to Brett's office and asked him to join me. He knew immediately as well. His stubbornness in the discussion surprised me. He challenged the decision at times, but I stuck with the script and didn't argue with him. I was a broken record with my mantra about the decision. I didn't waver. He eventually surrendered to the foregone conclusion.

Brett didn't shed a tear, not even close. Frankly, I expected Patrick to be the strong one and that Brett would have broken down. I guess what we know and see on the outside can hide a different interior. I had, over the last couple of weeks, questioned my decision about Brett. No more. It was clearly the right decision. And Patrick no longer pissed me off. I wish I'd known this more tender-hearted man the last couple of years.

We pulled the other 12 affected team members into a conference room and let them know as a group what was happening. Representatives from HR then met with each person individually and reviewed the severance packages in private. We had four HR reps helping, so we got through everyone pretty quickly. I sat in on the manager's discussions, thinking they may be more challenging communications. They weren't.

The team meeting went well. I hit most of the key points, missing a couple out of sheer nervousness. It felt as though I said 'um' a million times, but Jan said I did really well, one of the better ones she's experienced.

Most of what Carol told me was accurate about how the remaining staff would respond. I got a few questions about the people who were released—were they treated respectfully, did they get a good severance, will Cantril help them find new jobs. It was natural—people caring about their friends.

But most of the questions were about what to expect next.

They wanted to know what the new organization structure was going to look like. I was caught short. I should have predicted that question and either had the org chart ready or simply not mentioned it. Another lesson learned.

With a shaky voice, Jenny Fine asked why *those* individuals were let go. I was caught off-guard by the question and stuttered for a few seconds, searching for a way to answer the question without having to actually answer it. But then the words just came to me. I restated that they were good employees and contributed to IT and the business over their time with the company, but that Jim needed different results from IT and that meant different skills, abilities, talents and experiences. A new direction meant we needed new players, just like when a new general manager or manager takes over a professional sports team. New coaches and new players soon follow.

Everyone seemed to understand this. And I thought Jenny was brave to ask that question in such a caring manner, unafraid of putting me on the spot. I didn't know her well, except that she was a junior business analyst. I decided to follow up with her to see if she was someone we could use in bigger ways.

Then there was Cary Whitcanack's question. "Will there be a pizza party for the remaining staff?" It was the type of question Patrick would have asked. I just stared at him for a few seconds, my mouth slightly open. Maybe that number should have been 15! I think I finally managed to utter something like "We'll see," and quickly moved to the next question.

We were finished by 10:30 a.m. I wanted desperately to leave, to just go home to the comfort of my wife and kids. But I knew I needed to be there for my team, so I stayed.

* * *

I pulled into my driveway just before 7:00 p.m., turned off the engine and just sat.

I thought about each of the lives I'd affected that day. I'd been there myself and couldn't help but hurt for them. I was confident they would all land squarely on their feet, so my thoughts moved quickly to the next week… and then the following weeks and months. I began to feel excited for the first time, anticipating what I might be able to build in IT and what we might be able to accomplish for the business and our customers.

I had stolen some of Carol's words and made them my own. I told my team that I could clearly see the business cheering them on one day, one day very soon—congratulating us for winning a new customer to Cantril directly because of IT's leadership and collaboration with the business. I literally watched several people sit straight up in their chairs as if they were little kids hearing that Santa would be visiting the class that day.

That surprised me. They really were ready to engage and deliver at a higher level. I definitely sensed a greater excitement and energy in the room after that message. And I knew it was my job to harness that energy, point it in the right direction, and then set it free.

21

Setting IT Free

"What? Are you serious?" Steve asked.

I'd never seen him smile so big. He was always so focused and serious.

"Yes, Steve. I'd like for you to be my Director of Infrastructure, full-time. We've spent the last several weeks searching, and frankly, you're far and away the best candidate. Congratulations."

We did search for several weeks and had some good candidates, but each day, Steve grew more and more comfortable in the position he'd taken over temporarily after Patrick left. He immediately pulled himself out of the technical detail, prioritized what he wanted to fix and attacked the list. I was impressed. And it certainly didn't hurt that three of his team members approached me separately and said they'd like to see Steve be their manager permanently.

I told him that he was doing exactly what I needed him to be doing, and I encouraged him to keep it up. As he walked out of my office, I felt as though I had finally let go of the infrastructure group. I had set them free.

A couple weeks earlier, we had hired Brett's replacement as well. I'm not quite at the point of setting that new director free; I need to see him perform in his new role a while longer. I don't hover, and I stay out of his meetings. I'll learn soon enough if he's not a fit for us, but all indications are that he was a very good hire.

Kim rebuilt her team as planned. She hired Dominic Stewart to lead the PMO and promoted Alicia Martinez to lead the business alignment team. They rebuilt our Strategic Project Portfolio, got projects back on track, and made sure that all new projects followed our project management methodology.

And we renamed her team IT Strategy and put the business intelligence team in her group. Kim was rockin' and rollin'.

Jenny was promoted to senior business systems analyst and lead tactical alignment for both Sales and Marketing. Oh, and Cary got his pizza party. He was a happy little coder.

I clearly saw the impact of these changes across all of IT, and in our interactions with the business. I looked back and remembered the chaos that I had created under my unpredictable management style. What were once mediocre resources delivering mediocre results were now exceptional resources delivering unexpected results. I simply had to get out of my own way, get the right people in the right roles and set them free.

Setting my team free gave me the time to focus on my own priorities, those tasks that allow me to give the business the greatest return as CIO. I did pretty well staying out of the technical and tactical detail. Technology was always like a drug for me, though—one that kept calling out my name. But letting go got easier. Watching my team accomplish things on their own fulfilled me more than technology ever did.

Over the past few weeks, I learned that both Patrick and Brett landed new jobs. Patrick was making more money as a director for a cloud services provider. Brett landed as a director too, but with a much smaller company and much smaller staff. The developers all found jobs within a matter of days. Not surprising in this town. A couple of individuals were still looking, but I felt a little better about being so disruptive to people's careers and personal lives.

I walked into the office late one morning. The buzz in IT was nothing like I'd ever experienced before. All three conference rooms were in use. I saw SVPs in two of them and they actually looked engaged! We were executing business-IT alignment!

Rich, our SVP of operations, stuck his head in my door after he left his alignment meeting. "I don't know what you did, Brian, but thank you. IT is now listening, and that is a major accomplishment in this place. Keep up the good work!" He gave me a thumbs-up and walked out.

I just smiled. There was something more to it, though. Getting things done through other people added a dimension I'd never experienced before. Being able to sit back and watch the buzz of purposeful activity went beyond gratifying. I was impacting people's lives and careers, but in a wholly positive way.

Together we were accomplishing what we had envisioned and what we had communicated. Carol said there was a feeling to leadership. I understood it now. I felt it. I had the team I wanted. I felt on top of the world, or at least on top of the forest. I would have been happy to see it end there.

22

Winning Customers

I'm not much of a lunch eater. I keep crackers and other seemingly healthy foods in my desk and just snack all day. But when the CEO asks you to join him for lunch, you have to toss personal preferences aside. So, this morning, I kept the crackers in the drawer and worked up an appetite.

Jill asked me last night if I was afraid of what he might want to talk about. Funny question, actually. Given all that I'd been through over the past year, fear wasn't a feeling that entered my mind. I *knew* that IT was focused on the right things, that we were moving in the right direction, and that we were delivering higher-quality solutions on time and within budget. Jill wasn't convinced, though, and asked me to call her after lunch. Silly girl.

Jim loved sushi, so we headed to a place called Nara's. They have the absolute freshest seafood on the Eastside, he said. I grew up in Iowa, so for me to love sushi has been quite a surprise.

We had a quiet ride from the office to the restaurant. We spoke only sporadically, mostly about sports. I could tell something was on his mind.

We were seated immediately. We each ordered a selection of our favorites and agreed to share.

Jim jumped right into his serious business topic. "Brian, I wasn't confident that you would survive. I put a lot of pressure on you to respond quickly, but you did respond, eventually. I'm quite impressed."

"Thank you, Jim. That means a lot coming from you."

"Here, Brian. These are for you."

Jim handed me two envelopes. One was a long white standard-type envelope. My first name was hand-written on the front. The other was a large white envelope, one that would hold 8.5 x 11 documents. Cantril's logo was on the front. This one was formally addressed to me.

I reached out my hand and took them. I laid them on the table in front of me. I just stared at them, not sure whether to open them or wait. Jim saw that I was unsure of what to do.

"You don't have to open them now. The smaller envelope is from me. It's a gift card to a new restaurant in town I thought you and your family would enjoy."

I looked at him with a slight bit of fear. "What restaurant?" I asked, hoping and praying it wasn't Caye's House. It wasn't. Jill would have laughed, but I wouldn't have found it so funny.

"And the other envelope," Jim continued, "well, it's a stock option grant and a cash bonus for you. 3,500 shares. I'll let you look at the bonus in private."

I think my jaw dropped open. No, I *know* my jaw dropped open. I'd received bonuses before, but they were always expected. This one wasn't. I could tell Jim was pleased with my blatant non-verbal reaction. His smile grew larger.

'You earned it, Brian. Congratulations."

I sat in shock for a few seconds longer. Jim simply held his smile. That was the moment I knew. I knew what he was think-ing. He felt about me the same way I felt about Kim and Steve and Jenny and rest of my team. I played a small part in allowing Jim to lead, and lead well.

And in that moment I knew the pinnacle of leadership. I knew why Carol was helping me. She said I would figure it out on my own. I was in awe of this moment. I couldn't wait to tell Carol.

Jim went on to tell me that the client who called a few months earlier and told him that he would have to move to another distribution partner had called again. They chose to stick with Cantril Distribution after all. Jim was beaming.

I knew why. He didn't need to finish the story, but he did. After our IT re-org, we promoted Alexander Dixon to head up our business intelligence team. He wanted to build a BI vision, strategy and roadmap from scratch. We did start that initiative, but it was going to take months for it to produce results. The business, and our customers, needed something now.

So, he researched online BI solutions and found one that would have us up and running in just 30 days. This tool also allowed us to brand individual sites for each of our customers, which we did. The client was so impressed with how quickly we solved their most-pressing problem that they changed their mind about their distribution partner, even though they had already started the move.

"Brian, IT *can* drive business growth and profitability. Now that you've experienced it, even led it, I just need more." He smiled, almost laughed, as if he might be exaggerating his de-mand. But I knew he was serious. I was energized and excited to do more for him.

A year ago I would have tried to solve this and every other problem through my own efforts, not able to trust my team. But now I could relax knowing I had a team that could execute without me. Leaders get things done through people and, boy, did I have people, talented people who were excited to be making a difference for their company and its customers.

Without Jim waking me up a year ago and setting a higher expectation for IT, we would have permanently lost that client. And nothing in IT would have changed because I wouldn't have changed.

When I got back to my office I slowly opened the envelope and looked at the stock grant. Then I opened the check. I was floored by the amount. I called Jill and told her about the lunch, what Jim shared, and of course about the bonus. She got quiet.

"Are you okay, Jill?" I asked.

"I am, Brian. It's not the money. I'm just so proud of you, what you've accomplished for your company and what you've accomplished for yourself." I had to wipe away a tear hearing her words.

We finished our talk and hung up, both of us anxious to continue the conversation later that night. I can honestly say that I'd never felt closer to my wife than I did at that moment. I so much appreciated her love and support.

* * *

I called Carol later that afternoon and told her about my lunch with Jim. As we were talking, I got an idea. I shared it with Carol, and she loved it. I told her that I would take care of everything but needed her here in person tomorrow afternoon. We agreed to meet in my office at 1:00 p.m.

I called Lisa in and asked her to schedule a 30-minute meeting with Jim for 1:30 p.m. tomorrow, and to call me immediately if he wasn't available.

"What's the topic?" Lisa asked.

"It's a surprise, so make something up. You're smart and creative."

"You don't know what the meeting is about but you want me to set it up, is that right?"

I looked at Lisa blankly and realized then that I'd have to let her in on the secret. After I did, she got a gleam in her eye.

"Okay, I'll schedule the meeting, Brian, but I want in on it."

She was bribing me, so I teased her. "But you work for me. I order you to set up the meeting!"

"Nope," she responded.

"Lisa, you're disobeying your boss."

"I don't care…"

I smiled and gave in. "Okay, you're in."

"Yipeee!" she squealed.

Lisa called Jim's admin and set up the meeting. I headed back outside. I had a special errand to run.

The Pinnacle of Leadership

"Here it is, Carol. You're the last one to sign it."

Carol took the gift and card from me, grabbed a pen from my desk and signed her name. "What a great idea, Brian. When do we give it to him?"

"1:30."

Carol handed the gift and card back to me.

We had almost 30 minutes to kill before the meeting. I leaned against my desk. Carol sat in one of the chairs around the conference table. "Carol, do you remember mentioning pinnacle of leadership to me? I asked you then what it was and you refused to tell me. Remember?"

"I do, Brian. I felt bad not telling you, but I really did want you to discover it for yourself."

"Well, I think I did."

Carol's eyes widened. "And..."

"And I think the pinnacle of leadership is the joy of developing other leaders. Becoming a true leader is like a gift received, but once it's received, you can't help but pass it on. It's inherent in the traits of true leaders. It feels like a right-of-passage."

Carol nodded. "Well said, Brian. You are exactly right. How did you figure it out?"

"I saw it in Jim's eyes when he gave me my bonus. And I understood why you were willing to spend so much of your time with me. It's a joy for both of you to see people grow as leaders."

Carol smiled. "It's a pure joy, Brian, it really is."

"I'm not sure if I'm ready to do it myself, but I'm going to try. I may need to lean on you some as I venture down this new path with my team, helping them develop their own leadership abilities."

"You can certainly count on me, you know that."

I glanced at the time on my desk phone. We still had a few minutes before we headed upstairs. We talked about the past year. I shared a challenge that I was still experiencing in my role, and in true Carol fashion, she guided me through seeing the issue from different perspectives, and helped me discover potential solutions. It was 1:25 p.m., so we grabbed Lisa and headed upstairs to the boardroom.

As my management team and the rest of Jim's executive team walked into the conference room, I introduced them to Carol. Everyone was happy to meet her, commenting that Jim had spoken highly of her.

A couple minutes past 1:30 p.m., Jim walked in. He was taken aback by all of us just standing there in the conference room. "I thought this was a meeting about last quarter's numbers? Why are you all here? What's going on?"

His look grew more perplexed as we all just stood there smiling at him.

I broke the silence. "This *is* a meeting about the numbers," I said. "Tom, are you ready?"

"Yep. Jim, I have our final numbers for the previous quarter. Our revenue was up 12 percent over the same quarter a year ago and our EBITDA was up 19 percent, giving us two consecutive quarters of revenue and EBITDA growth, something this company hadn't achieved in several years."

Everyone cheered. The sales executives threw out a couple dog barks and fist waves.

Still perplexed, Jim said, "That's only two quarters. We still have plenty of work to do."

Someone unhappy with Jim's response shouted, "Party pooper!"

At that last comment, Jim surrendered to the moment. He relaxed his sturdy posture and broke a smile. "Okay, I give, what's up?"

I took the lead. "Jim, we appreciate what you've done for the company in such a short time. I especially want to thank you for kicking me in the butt a year ago and waking me up. And, of course, for introducing me to Carol. So here, this is for you."

I handed him the gift bag. He took it in his right hand and froze. He glanced down at the gift, then lifted his head and scanned the room catching everyone's eyes. I could tell he wasn't sure what to do.

I nodded my head at him. "Go ahead, open it."

He set the bag on the conference room table. He reached in and pulled out a square, clear plastic case, similar to the ones that held his father's baseballs. He looked at the empty case, and then looked inside the bag again.

"Hey Jim. This might fit in it." I tossed him a baseball signed by Carol, Lisa, his management team and my directors. He gazed at the ball for quite a long moment, slowly turning it in his hands to make sure he read every name. "It says 'MVP' on it, too. Did you see that?" I asked.

Jim looked up. "Thank you everyone. This is quite unexpected but also very touching. This is a special gift. It brings together many of the things I care most about—my dad, baseball, business and this team. Thank you. I'm sure you all know where this ball will go."

He placed the ball inside the case. "I would like to know who wrote 'MVP' on the ball. It couldn't have been you, Brian, because I doubt that an IT guy could spell MVP."

My eyes shot wide open and I jerked my head toward Carol. "You told him!" She covered her mouth with one hand to keep from laughing out loud and held her other hand up so I could see that her fingers were crossed.

"That'll teach you to make fun of a CEO!" Jim said, laughing. I had to laugh too. We all spent the next few minutes taking turns looking at the baseball. Carol hugged Jim goodbye and excused herself. Everyone else trickled out. Jim and I were the last to leave.

"Jim, do you remember the story you told me in your office about Danny and his new glove?"

"Of course, Brian."

"Did your dad buy that glove for him and put it on his porch that morning?"

Jim looked down and paused for a few seconds. I again watched him go back to those days playing ball on his dad's teams. Still lost in thought, Jim nodded. "He did, Brian. He told me that 20 years after it happened. My dad said that he couldn't afford to buy that new glove, but he knew it would make Danny happy, and that was worth more than the cost of the glove to him.

"He touched Danny's heart that day, I know that. He connected with him in a way that was meaningful to both of them. He did that for most of the boys he coached, some in small ways, some in big ways. Those connections built teams that won and lost together, versus a group of boys that happened to wear the same uniform."

"Why did it take him 20 years to tell you?"

"My dad was a quiet man. He didn't talk much. His actions spoke for his heart—and revealed his heart."

"Jim, from where I'm standing, you build winning teams the same way your dad did."

"Quiet, Brian, or you're going to make an old man cry."

I looked Jim in the eyes and smiled. He slapped me on the back, and we headed for the door together. As we parted in the hallway, he called back to me.

"Oh, Brian. Thank you."

We had already said thank you a hundred times over the last hour. But this one felt different.

"What for?"

He grinned. "I got a good night's sleep last night."

Those first talks with Jim, a year old now, raced through my mind. I remembered the stories of his dad, his little league teams and that CIO who haunted his CEO. My own feelings of fear, inadequacy and doubt, so intertwined with Jim's stories, rushed through me as well. And then they disappeared.

I pulled my thoughts back to Jim and smiled.

"I did too, Jim. I did too."

The first person you lead is you.

—John Maxwell

Epilogue

Leadership is visionary and trusting and freeing. It's amazing how people respond when lead, truly lead.

And as Brian learned, it is possible to build an IT organization, any organization for that matter, that is in sync with rest of the business—that helps drive greater business growth, profitability and even innovation. True leadership achieves this and sustains it over time.

It is very difficult to make this transition though. Not impossible, but difficult, as Brian discovered. The key is to have a vision for yourself as a leader, and to pursue that vision each and every day. And when you allow yourself to learn from your inevitable missteps and failures, you will actually shorten the path to true leadership.

A mentor or executive coach can also help streamline and shorten this path, while reducing frustration along the way. There's a reason the very best athletes in the world have coaches, as do the very best business leaders.

And it is impossible to acquire leadership abilities just by reading a book (yes, even this book), or even by attending a training class. It's no different than reading the top 10 tips for a better swing in a golf magazine. Until you put the article down, pick up your clubs and practice, it's of no value. Leadership, too, requires development. Strike that. Leadership requires ongoing, continual development—never-ending rounds of practice.

So pick one or two concepts in the book that you desire to develop or improve, set measurable goals, write them down, tell someone who can help hold you accountable, and then start taking daily action toward those goals. As you perfect one, introduce another in the same fashion. Then repeatedly repeat.

Finally, the chart on the next page allows you to quickly gauge the value of your IT organization within your company. Increasing IT's value to the business (and to its customers/clients) requires continually improving IT's efficiency, effectiveness and leadership abilities.

To use the chart, ask yourself the following three questions:

1. Where is our IT organization today?

2. Where would I like it to be?

3. Is it moving in the right direction?

If your answer to #3 was no, my desire is that you now see true leadership as the solution, the solution for positioning IT to deliver greater business value—greater growth, profitability and innovation.

IT Value Continuum
© GrowthWave

Continual improvement in efficiency, effectiveness and leadership →

1	2	3	4	5
Disrupts/ distracts the business	Operates as a utility, meets technical needs	Improves business productivity	Propels strategic growth	Drives innovation

Your IT investments deliver:

C O S T S

R E T U R N S

Stronger business alignment →